EDUCATION IN A COMPETITIVE AND GLOBALIZING WORLD

# HIGHER EDUCATION OF PEOPLE OF COLOR

## VIEWS ON EFFECTIVENESS OF HISTORICALLY BLACK COLLEGES AND UNIVERSITIES AND ENCOURAGING PURSUIT OF STEM CAREERS

# EDUCATION IN A COMPETITIVE AND GLOBALIZING WORLD

Additional books in this series can be found on Nova's website under the Series tab.

Additional e-books in this series can be found on Nova's website under the e-book tab.

EDUCATION IN A COMPETITIVE AND GLOBALIZING WORLD

# HIGHER EDUCATION OF PEOPLE OF COLOR

## VIEWS ON EFFECTIVENESS OF HISTORICALLY BLACK COLLEGES AND UNIVERSITIES AND ENCOURAGING PURSUIT OF STEM CAREERS

DANA G. PERRY
EDITOR

nova publishers
*New York*

**Library of Congress Cataloging-in-Publication Data**

ISBN: 978-1-63483-525-1

*Published by Nova Science Publishers, Inc. † New York*

# CONTENTS

# PREFACE

The Black College and University Act defined an historically black college and university (HBCU) as one that existed before 1964 with a historic and contemporary mission of educating blacks while being open to all. An HBCU must either have earned accreditation from a nationally recognized accrediting agency or association or be making reasonable progress toward accreditation. The U.S. Commission on Civil Rights conducted a briefing on May 5, 2006, to assess the educational effectiveness of HBCUs. This book discusses HBCUs and examines why minority college students who begin their college studies intending to major in science, technology, engineering or math (STEM) leave these disciplines in disproportionate numbers before graduation.

In: Higher Education of People of Color
Editor: Dana G. Perry
ISBN: 978-1-63483-525-1


*Chapter 1*

# THE EDUCATIONAL EFFECTIVENESS OF HISTORICALLY BLACK COLLEGES AND UNIVERSITIES*

*U.S. Commission on Civil Rights*

**Letter of Transmittal**

The President
The President of the Senate
The Speaker of the House

Sirs and Madam:

The United States Commission on Civil Rights (Commission) is pleased to transmit this report, *The Educational Effectiveness of Historically Black Colleges and Universities.* A panel of distinguished experts briefed members of the Commission on the educational effectiveness of these institutions, which have been pivotal in educating students, especially black students, for generations. Among the topics addressed was how adequately these institutions prepared students for the increasing demand for highly skilled workers. Based on that briefing, the Commission developed the findings and recommendations that are included in this report.

According to the data presented, the Commission found that students at historically black colleges and universities ("HBCUs") reported higher levels of academic engagement on some survey dimensions than their counterparts at non-HBCUs and that black students at HBCUs were more likely to be involved in faculty research projects than black students at non-HBCUs. HBCUs also produce a disproportionately high share of black students who receive degrees in science, engineering, technology, or mathematics (the "STEM" fields), the Commission found. The Commission attributed the HBCUs' success in educating and graduating disproportionately large numbers of black students (both generally and with

* This is an edited, reformatted and augmented version of a briefing report issued November 2010.

respect to those graduating with STEM degrees) in part to their admission policies, which do not create the situation of academic mismatch (a credentials gap) often found at non-HBCUs. Mismatch occurs frequently in institutions that grant preferential admissions based on race. The Commission credited mismatch, rather than discrimination or black students' alleged lack of interest in science, with HBCUs' comparative success over non-HBCUs in producing a higher proportion of black STEM graduates.

The Commission recommended that black students interested in STEM majors may wish to consider attending a college or a university, including an HBCU, at which their academic credentials match those of the typical student so that they avoid experiencing the negative effects academic mismatch. The Commission also urged state governments to allocate funds to HBCUs and non-HBCUs based on racially neutral criteria. Finally, the Commissioner asked that researchers carefully study the success of HBCUs to learn how other schools can emulate their best practices.

Part A, which consists of the body of this report, was approved on March 12, 2010 by Chairman Reynolds, Vice Chair Thernstrom, and Commissioners Heriot, Kirsanow, Melendez, and Yaki. Commissioner Gaziano abstained. Vote tallies for each of the Commission's findings and recommendations, which make up Part B of the report, are noted therein.

For the Commissioners,

Gerald A. Reynolds
Chairman

## EXECUTIVE SUMMARY

The Black College and University Act defined an historically black college and university (HBCU)[1] as one that existed before 1964 with a historic and contemporary mission of educating blacks while being open to all. An HBCU must either have earned accreditation from a nationally recognized accrediting agency or association or be making reasonable progress toward accreditation. Currently, 103 HBCUs are located mainly in the Southeastern United States, the District of Columbia, and the Virgin Islands.

The U.S. Commission on Civil Rights conducted a briefing on May 5, 2006, to assess the educational effectiveness of HBCUs. The Commission invited five distinguished panelists to discuss the issue: Louis W. Sullivan, founding dean and first president of Morehouse School of Medicine, as well as a presidential advisor and former cabinet secretary; Earl S. Richardson, president of Morgan State University and a former presidential advisor on Historically Black Colleges and Universities; Jamie P. Merisotis, president of the Institute for Higher Education Policy; Raymond C. Pierce, dean and professor of law at North Carolina Central University and a former deputy assistant secretary at the U.S. Department of Education; and Mikyong Minsun Kim, associate professor at the George Washington University's Graduate School of Education and Human Development.

After the briefing, Commissioners offered two articles to help provide the reader with a richer understanding of the subject because policymakers should base conclusions about the efficacy of HBCUs in educating black students as compared to non-HBCUs on stronger evidence than mere public support. In both studies the researchers rely on extensive empirical data to reach their conclusions. The two articles are reproduced at the end of this report and cast some additional light on the briefing topic.

In one study,[2] economists Fryer and Greenstone found that during the 1970s, HBCU attendance resulted in a greater likelihood of graduation when compared to black attendees at traditionally white institutions, as well as an increased probability of majoring in a physical science. As important, HBCUs provided a substantial wage premium to their graduates. However, by the 1990s, HBCU attendance yielded a substantial wage penalty to black matriculants in comparison with black students attending a traditionally white institution.

Other differences are more positive. Students attending HBCUs appear to demonstrate increased charitable giving, political participation, religious participation, and propensity to major in the physical sciences compared with those who went to traditionally white institutions.

In the second research effort Commissioners identified, psychologists Elliott, et al.,[3] sought to assess the effects of black students attending HBCUs rather than non-HBCUs. The study focused on predictors of whether or not black students decide to major in the sciences with some discussion of the special role HBCUs might play in this process. The authors note an initial substantial interest in, and favorable feelings toward, science among blacks pursing science careers, an effect more positive than that found among white students. The black students failed to manifest such interest later in college and subsequently. To understand this change, the researchers studied the likelihood of students enrolled in four Ivy League institutions majoring in science and, to a lesser extent, of their dropping out of college completely. Elliott, et al., were particularly interested in the situation of high-ability minority students.

The researchers' overall finding is that "preadmission variables accounted for a significant fraction of the variance in persistence decisions, while ethnicity did not." Although black students have a greater initial interest in choosing science majors and careers than whites, and an attraction equal to that of all nonblacks, the heightened appeal of these pursuits is not sustained. While students' initial interest predicts their persistence in the field of science, the enhanced attraction does not stem the disproportionately large attrition among black students. The analysis indicates that the greater attrition of black majors arises from a very large disadvantage in black students' average developed academic ability compared to whites and Asian Americans.

Elliott, et al., indicate that the *relative* position of blacks in HBCUs as compared to those at non- HBCUs probably accounts for the positive effects of HBCUs. The authors note that black students at non-HBCUs—even those who score well compared to national racial and ethnic norms—are competitively disadvantaged relative to their school's student body on indicators of developed ability that predict science persistence and achievement, particularly at elite colleges and universities. This results in weaker performance and persistence among blacks than might occur in less competitive settings. The authors note that the differences are large enough to suggest that the achievement gaps are due to nonacademic factors as well as to differences in developed ability.

Most importantly in the context of this briefing, Elliott, et al., note that while HBCU enrollees and graduates have quite low SAT scores and high school grades, they nonetheless produce 40 percent of black engineers with only 20 percent of black enrollment. They also note that the top 21 undergraduate producers of blacks with doctoral degrees were HBCUs, and none of the highly successful schools were among the 30 most selective academic institutions.

Elliott, et al., found support for their comparative performance hypothesis in a re-analysis of data from 11 private colleges. The research shows that, when students are ranked by SAT math scores and divided into three groups, each institution awarded science degrees to, on average, 54 percent of students in the top tercile, 31 percent in the middle group, and only 15 percent in the bottom third, regardless of differences across colleges in each tier's average math scores. While having a higher SAT math score is associated with becoming a science graduate, the rank of an SAT math score within any given institution also matters in predicting whether a student will become a science graduate. For example, a student among the top third of a less selective institution is likely to have a far greater probability of graduating with a science degree than a student in the bottom tercile of a selective institution, even when they have the same SAT math score. The example that Elliott and his coauthors provide predicts that 55 percent of the former, and only 15.4 percent of the latter, are likely to graduate with a science degree. Relative student position in institutions matters. Here, then, the HBCUs can continue to have an important role in nurturing academic talent among black students.

Despite widely ranging methods, data sources, and professional disciplines, theories, and analyses, both studies that Commissioners contributed to the record after the briefing support the conclusion that HBCUs make a distinctive contribution to the well-being of black Americans.

In the briefing session, Dr. Sullivan stated that for some young African-Americans the development that occurs at HBCUs might have a profound influence on their lives. For example, African-American graduates of certain HBCUs, such as Xavier University, Spelman College, and Morehouse College, successfully gained entry into graduate, medical, engineering, law schools and other fields, in percentages and numbers equaling or exceeding those of African-American students that attended wealthier, longer established, and predominantly-white institutions. He noted that graduates of the Morehouse School of Medicine pass national medical examinations at rates equaling or surpassing the overall national pass rate even though the institution is one of the most recently established medical schools in the United States. Dr. Sullivan suggested that faculty members' dedication to teaching, a supportive social environment on campus, faculty and staff members' encouragement of students to explore leadership roles in their chosen careers, and the general availability of faculty role models help explain HBCU success.

Dr. Richardson indicated that after the enactment of the Civil Rights Act of 1964 (CRA) and initial federal enforcement in the early 1970s, questions arose about the future role of HBCUs in contemporary higher education. Previously, there was a clear understanding that HBCUs were open to all students and were the only type of institution in the Southern states available to African-American students seeking a college education. Continuing, he explained that the CRA mandated that non-HBCUs increase minority enrollment but did not require HBCUs to raise their white enrollment. He went on to say that African-American educational attainment seriously lagged behind that of whites despite HBCUs' success in enabling access, and that the gap between the two groups has been widening in recent years. He observed that HBCUs are far more productive in graduating African-Americans than are selective institutions and therefore must be an "integral part of any strategy for increasing the number of black graduates in the nation." Elliott, et al., made a comparable point on HBCU productivity, albeit pertaining to the graduation of science graduates.[4]

Mr. Merisotis said that a 2004–2005 national survey of student engagement found that, compared to African-American students enrolled in "predominantly white institutions," those attending HBCUs reported more interactions with faculty members. He indicated that the 2004 Campus Compact[5] membership survey documented that HBCUs are more likely to require service and service-learning as a condition of graduation. Mr. Merisotis also said HBCUs provide "additional support, guidance, and mentoring to students from educationally and economically disadvantaged backgrounds and disabled students to ensure their admission into and graduation from college. Finally, he drew attention to the success of the undergraduate-level "HBCU UP" program that the National Science Foundation (NSF) funds at 14 HBCU sites. NSF, he said, reported that at HBCUs in which the program has been in place for five years, the passing rates in required mathematics courses improved as did performance in mandatory science, technology, and engineering courses. Of the science, technology, engineering, and mathematics (STEM) graduates from these HBCUs, he said about 25 percent have acquired undergraduate research experience that would be helpful to them in graduate school. It bears noting that existing empirical literature suggested that blacks aspire to be in science.[6] Elliott, et al., pointed out that "intention to concentrate in science is by far the strongest predictor of actually doing so."[7]

Dean Pierce stated that HBCUs continue to educate large numbers of African-Americans effectively, thus contributing to the nation's need for a learned population and skilled workforce. He indicated that desegregation efforts in state higher education systems involved private lawsuits seeking equal funding for HBCUs relative to traditionally white institutions in conjunction with federal government intervention. The 1970s *Adams* cases are illustrative.[8] In these cases, Dean Pierce said private individuals filed suit in federal court against the Secretary of the U. S. Department for Health, Education, and Welfare (HEW), the agency itself, and its Office for Civil Rights (OCR) for failure to enforce federal civil rights laws with regard to African-Americans enrolled in HBCUs. The court found against OCR and directed it to develop guidelines to desegregate state systems of higher education. OCR's 1978 policy, he explained, examined two components to determine civil rights compliance—HBCU enhancement and traditionally white institution affirmative action in recruitment to attract and enroll more African-American students. Following *Ayers v. Fordice,* 1992 (*Fordice),* he said OCR created a policy for higher education desegregation in 1994 that was far more rigorous than that of the 1978 one. Continuing, he offered that the "problem we face today is almost a re-visitation of the situation that led to the *Adams* cases in the 1970s." In 2000, 2001, and especially 2002, he observed that some states were seeking to revert to a policy of hindering HBCUs' ability to offer attractive educational programs, while others were unnecessarily duplicating programs.[9] Historically, he said program duplication served to create and maintain a segregated higher education system.

In her turn to speak, Dr. Kim first provided some characteristics about HBCUs and their students. Compared to "historically white colleges and universities," she said that HBCUs tend to have fewer resources, smaller enrollments, a lower student-faculty ratio, and higher student-faculty interactions. Of their students, she said that they were close to 1.5 times likelier than students at historically white colleges and universities to participate in faculty members' research, their family backgrounds tended to be less affluent, and they are generally less academically prepared. With respect to the latter, the research of Elliott, et al., noted, for example, that "black grade-12 achievement in math is about the same as, and in science a little worse than, white grade-8 achievement."[10] Dr. Kim then discussed her research, which examined the impact of HBCUs and historically white colleges and universities on three academic outcomes: overall academic ability, writing ability, and mathematics ability. She found no significant difference between African-American students at HBCUs and historically white colleges and universities in these three areas. Similarly, she found no significant difference between African-American students in HBCUs and historically white colleges and universities in the likelihood of earning a baccalaureate degree and in early career earnings.[11]

Dr. Kim pointed out that these findings showed that an African-American student's attendance at an HBCU or an historically white college or university did not affect his or her academic success and early career earnings. She said that her findings along with those of others "lend support to the proposition that HBCUs contribute significantly to higher education in this country and merit strong support."

The discussion began with Commissioner Kirsanow commenting on Dr. Kim's research. He noted that Dr. Kim did not control for regional effects in her study. Nonetheless, given HBCUs lower funding levels and the academic under-preparedness of their students on admission, such institutions were "doing a much better job than historically white colleges and universities in educating African-American students," he said, more than the findings of no significant difference suggest. Dr. Sullivan stated that Spelman College's graduation rate of 77 percent exceeds those of several respected and well-resourced white institutions. He added that Morehouse School of Medicine students performed better on the United States' Medical Licensing Examination than the country's other medical students despite scoring lower on SATs. Dean Pierce pointed out that in 2005, despite fewer resources and pre-entry academic profiles predicting lower achievement, graduates of North Carolina Central University School of Law achieved a first-time bar passage rate of 81 percent, which tied with that of Duke University School of Law. Vice Chair Thernstrom expressed admiration for HBCUs. During a visit to Savannah State University some years ago, she noted that many of its African-American students have limited academic skills, a reflection of the "racial gap in academic achievement at the end of high school." She was impressed by the quality education the institution offered and its dedicated efforts to improve students academically.

Commissioner Kirsanow found it striking that HBCUs awarded 40 percent of all black graduate STEM degrees. Further, he observed that the top 30 HBCUs graduated the largest numbers of blacks who enrolled in STEM doctoral programs. In the same vein, the Elliott, et al., article reported that despite generally weak pre-college academic credentials, HBCUs produced 40 percent of black science and engineering degrees even though they enrolled only 20 percent of the total black undergraduate students.[12] Elliott and his coauthors also commented that of the top 21 undergraduate producers of black doctoral graduates during the period 1986–1993, 17 were HBCUs.[13]

Panelists offered several suggestions for HBCUs' educational success with students whose precollege profiles are not sterling. Dr. Sullivan suggested that faculty commitment played a role.

Dr. Richardson and Dean Pierce both spoke of HBCUs' nurturing environment wherein a culture of engaging students within and outside the classroom flourishes. Both Dr. Richardson and Dr. Kim cited the importance of role modeling. Of interest here is the finding of Elliott, et al. that persistence is not just a matter of average preparation but of comparative position as well. A reasonably well-prepared student at an HBCU is in a strong competitive position in his or her institution, unlike a similar student at an elite, majority white institution, and is thus more likely to succeed and persist beyond the baccalaureate.[14]

Chairman Reynolds noted the severity of the budget reality and HBCUs' heavy dependence on state and federal funding, and then asked if HBCU officials have considered any alternative funding model. Dr. Richardson responded that in fiscally difficult times, the higher education community has to be particularly efficient in delivering quality education. He proposed a complementary system of higher education that strategically locates high quality academic programs aimed at developing the state's work force in a few geographically dispersed institutions, rather than unnecessarily duplicating them.

***At the close of the discussion, Commissioner Kirsanow asked panelists to identify key federal policy initiatives that could sustain and enhance HBCUs and significant threats that could imperil them. With respect to the former, panelists proposed a return to complementary academic programming among institutions within a higher education system while avoiding unnecessary program duplication in geographically proximate HBCUs and non-HBCUs, continued enhancement ofHBCUs, enforcement offederal civil rights laws, additional Title III supportfor HBCUs, and an increase in financial aid. The threats that panelists identified are unnecessary program duplication, declining financial resources, and public challenges to HBCUs' right to exist.***

Based on panelist presentations and the ensuing discussion during the briefing, the Commission offers the following findings and recommendations:

# Findings and Recommendations

## Findings

1) A historically black college or university ("HBCU") is a college or university that existed before 1964 that has a historic and contemporary mission of educating African-Americans while being open to all students. There are currently 103 HBCUs in the United States. Forty are public four-year colleges and universities, and 49 are private. The remaining 14 are two-year colleges, 11 of which are public and three private [Chairman Reynolds and Commissioners Gaziano, Heriot, Kirsanow, and Taylor voted in favor of this finding. Commissioner Yaki abstained.]
2) Many distinguished and highly successful Americans have graduated from one of these institutions, including high percentages of African-American congressmen, professors, CEOs, lawyers, and judges. Indeed, during the days of legally mandated segregation, academically talented African-American students desiring university degrees often had few or no options other than HBCUs. Today's African-American students have many more choices, among them universities formerly segregated by law or custom, leading some commentators to question what role HBCUs should play now and in the future [Chairman Reynolds and Commissioners Gaziano, Heriot, and Kirsanow voted in favor of this finding. Commissioners Taylor and Yaki abstained.]

3) According to survey data collected by the National Study for Student Engagement (NSSE), students at historically black colleges and universities ("HBCUs") report higher levels of engagement on some survey dimensions than do their counterparts at non-HBCUs. For example, according to 2004 and 2005 NSSE data from 37 HBCUs, African-American students report more contact with faculty than African-American students at non-HBCUs. Other studies also show that African-American students at HBCUs are more likely to be involved in faculty research projects than are African-American students at non-HBCUs [Chairman Reynolds and Commissioners Gaziano, Heriot, Kirsanow, and Taylor voted in favor of this finding. Commissioner Yaki abstained].
4) Generally, HBCUs have less funding and fewer support resources for their students than comparable non-HBCUs. A study conducted by one panelist, Dr. Kim, nonetheless found no significant differences in academic outcomes between HBCUs and wealthier non-HBCUs [Chairman Reynolds and Commissioners Gaziano, Heriot, Kirsanow, and Taylor voted in favor of this finding. Commissioner Yaki abstained].
5) HBCUs have an average graduation rate of 55 percent, which is lower than the 63 percent average graduation rate for non- HBCUs. This may occur, in part, because of HBCUs' generally liberal admission policies and their average student's weaker academic profile. However, for similarly situated students, attendance at an HBCU versus a non-HBCU has no differential effect on an individual African-American student's chances of obtaining a bachelor's degree [Chairman Reynolds and Commissioners Gaziano, Heriot, Kirsanow, and Taylor voted in favor of this finding. Commissioner Yaki voted against this finding].
6) HBCUs succeed in educating and graduating disproportionately large numbers of African-American students in part because their admission policies do not create the situation of academic mismatch often found at non-HBCUs. Many African-American students granted preferential admission at elite non-HBCUs, even when they score well compared to national norms, are competitively disadvantaged in developed ability relative to their school's student body who are admitted without consideration of racial or ethnic preferences. Thus, at some of these institutions, academically well-prepared non-Asian minority students, including African-Americans, have weaker performance and persistence rates than might occur in settings where the competition is in line with their current academic preparation [Chairman Reynolds and Commissioners Gaziano, Heriot, and Kirsanow voted in favor of this finding. Commissioner Yaki voted against this finding. Commissioner Taylor abstained].
7) HBCUs also produce a disproportionately high share of African-American students who receive degrees in science, engineering, technology, or mathematics (the "STEM" fields). Though only about 20 percent of African-American college students attend HBCUs, 40 percent of all African-American engineers received their degrees from an HBCU. Similarly, of the top 21 undergraduate producers of African-American science PhDs, 17 were HBCUs. The prevalence of academic mismatch, caused by non-HBCUs granting preferential admission to certain minority students as opposed to overt discrimination against African-Americans at non-HBCUs, or African-American students' lack of interest in science, appears to best explain HBCUs' successes in producing African-American STEM graduates

[Commissioners Gaziano, Heriot, Kirsanow, and Taylor voted in favor of this finding. Chairman Reynolds and Commissioner Yaki abstained].

## Recommendations

1) African-American students interested in STEM majors may also particularly wish to consider attending a college or a university, including an HBCU, at which their academic credentials match those of the typical student so that they avoid experiencing the negative effects of academic mismatch [Chairman Reynolds and Commissioners Gaziano, Heriot, and Kirsanow voted in favor of this recommendation. Commissioner Yaki voted against this recommendation. Commissioner Taylor abstained].
2) State governments should allocate funds to HBCUs and non-HBCUs based on racially neutral criteria [Chairman Reynolds and Commissioners Gaziano, Heriot, and Kirsanow voted in favor of this recommendation. Commissioner Yaki voted against this recommendation. Vice Chair Thernstrom and Commissioner Taylor abstained].
3) Researchers should carefully study the success of HBCUs to learn how other schools can emulate their best practices. [Chairman Reynolds and Commissioners Gaziano, Heriot, Kirsanow, and Taylor voted in favor of this recommendation Vice Chair Thernstrom and Commissioner Yaki abstained].

# INTRODUCTION

Section 322 of Title III of the Black College and University Act offers criteria that define historically black colleges and universities (HBCUs). An HBCU is a college or university that existed before 1964, with a historic and contemporary mission of educating blacks[15] while being open to all.[16] An HBCU must either have earned accreditation from a nationally recognized accrediting agency or association, or be making reasonable progress toward accreditation.[17] This definition covers a wide range of HBCUs—private, public, church-affiliated, large, small, rich, and poor.[18] Currently, 103 HBCUs are located mainly in the Southeastern United States, the District of Columbia, and the Virgin Islands. Of these, 40 are public four-year colleges or universities and 49 are private-four-year ones. The remaining 14 are two-year colleges, 11 of which are public and three private.[19]

HBCUs were created to fill a gap in the higher education of black students prior to the civil rights era. The historic mission of HBCUs grew out of racially discriminatory policies in education that were emblematic of an era of segregation in the broader society. Before the Civil War (1861-1865), public policy in the South prohibited the education of blacks,[20] a majority of whom were enslaved.[21] A handful of freed blacks, however, found rare opportunities to attend white colleges in the North.[22] After the Civil War, Christian missionaries established colleges in some Northern states to educate free blacks or runaway slaves. The earliest missionary-established historically black colleges were Cheney University, founded in 1837, Lincoln University, founded in 1854, both located in

Pennsylvania, and Wilberforce University in Ohio, founded in 1856.[23] Most private HBCUs came into existence during this period.[24]

The first Morrill Act of 1862 indirectly contributed to the founding of public HBCUs. Under this law, each state received grants of federal land to build colleges to teach agriculture, mechanical arts, and liberal education. Many states, however, prohibited blacks from attending white colleges.[25] The second Morrill Act of 1890 mandated states to make available separate educational facilities for blacks or admit them to white colleges.[26] Southern states chose to set up separate public colleges for blacks rather than admit them to white colleges.[27]

While *de jure* segregation is a thing of the past, HBCUs still enroll a substantial proportion of black students. In the pre-civil rights era, a large majority of blacks who attended colleges enrolled in HBCUs. Currently, about 20 percent of black students attend four-year HBCUs.[28] More strikingly, even today, HBCUs have educated a large proportion of black leaders including 40 percent of all Congressmen, 50 percent of non-HCBU professors, 12.5 percent of chief executive officers, 50 percent of lawyers, and 80 percent of judges.[29]

The Supreme Court's decision in *Ayers v. Fordice*[30] made the situation of HBCUs more precarious. *Fordice* instructed state legislatures either to find an educational justification for the continued existence of HBCUs or to integrate them with the non-HBCU institutions of their respective states. However, public support for HBCUs remains strong. As part of the recently passed College Cost Reduction Act, Congress increased the federal investment in HBCUs by over $170 million over the next two years.[31]

The U.S. Commission on Civil Rights (hereafter the Commission) conducted a briefing on May 5, 2006, to assess the educational effectiveness of HBCUs.[32] These institutions, despite difficult beginnings, have been pivotal in educating students, especially African-Americans, for generations. The Commission consulted numerous experts on HBCUs to identify distinguished panelists to discuss this important issue. The five panelists were:

- Louis W. Sullivan, chair of the President's Board of Advisors on Historically Black Colleges and Universities, founding dean and first president of Morehouse School of Medicine, and former Secretary of the U.S. Department of Health and Human Services;
- Earl S. Richardson, president of Morgan State University and former chair of the President's Board of Advisors on Historically Black Colleges and Universities;
- Jamie P. Merisotis, president of the Institute for Higher Education Policy;
- Raymond C. Pierce, dean and professor of law at North Carolina Central University, and former deputy assistant secretary for civil rights at the U. S. Department of Education; and
- Mikyong Minsun Kim, associate professor of Higher Education at The George Washington University's Graduate School of Education and Human Development and director of the Higher Education Administration Doctoral Program at the institution's Virginia Campus.

After the briefing, Commissioners offered two articles to help provide the reader with a richer understanding of the subject because policy makers should base conclusions about the efficacy of HBCUs in educating black students as compared to non-HBCUs[33] on stronger evidence than mere public support. In both studies the researchers rely on extensive empirical data to draw their conclusions. The two articles are reproduced at the end of this report and cast some additional light on the briefing topic.

In one study, economists Fryer and Greenstone analyzed several National Center for Education Statistics' databases collected in the 1970s and the 1990s. They found that during the 1970s, HBCU attendance resulted in a greater likelihood of graduation when compared to black attendees at traditionally white institutions, as well as an increased probability of majoring in a physical science.[34] As important, HBCUs provided a substantial wage premium to their graduates. However, by the 1990s, HBCU attendance yielded a substantial wage penalty to black matriculants in comparison with black students attending a traditionally white institution.[35]

Other differences are more positive. Students attending HBCUs appear to demonstrate increased charitable giving, political participation, religious participation, and propensity to major in the physical sciences compared with those who went to traditionally white institutions.[36]

If these conclusions survive further testing, they indicate that attending an HBCU no longer provides the economic advantage to black matriculants that it once did. It may also imply, as Fryer and Greenstone suggest, that traditionally white institutions may be more efficacious in placing their black matriculants than HBCUs. However, HBCUs' wage penalties are offset at least partly with non-pecuniary advantages that may accrue to black matriculants in attending HBCUs rather than traditionally white institutions. These include greater civic participation and greater willingness of students to major in the physical sciences. The large proportions of black lawyers, judges, members of the Congressional Black Caucus, and professors graduated from HBCUs, mentioned earlier, support such a finding.

In the other research effort Commissioners identified, psychologists Elliott, et al., sought to assess the effects of black students attending HBCUs rather than non-HBCUs. The study focused on predictors of whether or not black students decide to major in the sciences with some discussion of the special role HBCUs might play in this process. The authors note an initial substantial interest in, and favorable feelings toward, science among blacks pursing science careers, an effect more positive than that found among white students. The black students failed to manifest such interest later in college and subsequently.[37] To understand this change, the researchers studied the likelihood of students enrolled in four Ivy-league institutions majoring in science and, to a lesser extent, of their dropping out of college completely.[38] Elliott, et al., were particularly interested in the situation of high-ability minority students. Their study examined data from 1988 to1992.[39]

The researchers' overall finding is that "preadmission variables accounted for a significant fraction of the variance in persistence decisions, while ethnicity did not."[40] Although black students have a greater initial interest in choosing science majors and careers than whites, and an attraction equal to that of non-blacks, the heightened appeal of these pursuits is not sustained. Students' initial interest ordinarily does predict their persistence in the field of science. This study, however, finds that the enhanced attraction does not stem disproportionately large attrition among black students majoring and pursuing careers in science. The analysis indicates that the greater attrition of black majors arises from a very large disadvantage in black students' average developed academic ability compared to whites and Asian Americans.[41]

Elliott, et al., indicate that the *relative* position of blacks in HBCUs as compared to those at majority white institutions probably accounts for the positive effects of HBCUs. The authors note that black students at majority white institutions—even those who score well compared to national racial and ethnic norms—are competitively disadvantaged relative to their school's student body on indicators of developed ability that predict science persistence and achievement, particularly at elite colleges and universities. This results in weaker performance and persistence among blacks than might occur in less competitive settings.[42] The authors note that the differences are large enough to suggest that the achievement gaps are due to nonacademic factors as well as to differences in developed ability.

Most importantly in the context of this briefing, Elliott, et al., note that while HBCU enrollees and graduates have quite low SAT[43] scores and high school grades, they nonetheless produce 40 percent of black engineers with only 20 percent of black enrollment.[44] They also note that the top 21 undergraduate producers of blacks with doctoral degrees were HBCUs, and none of the highly successful schools were among the 30 most selective academic institutions.[45]

Elliott, et al., found support for their comparative performance hypothesis in a re-analysis of data from 11 private colleges. The research shows that, when students are ranked by SAT math scores and divided into three groups, each institution awarded science degrees to, on average, 54 percent of students in the top tercile, 31 percent in the middle group, and only 15 percent in the bottom third, regardless of differences across colleges in each tier's average math scores. While having a higher SAT math score is associated with becoming a science graduate, the rank of an SAT math score within any given institution also matters in predicting whether a student will become a science graduate. For example, a student among the top third of a less selective institution is likely to have a far greater probability of graduating with a science degree than a student in the bottom tercile of a selective institution, even when they have the same SAT math score. The example that Elliott and his coauthors provide predicts that 55 percent of the former, and only 15.4 percent of the latter, are likely to graduate with a science degree.[46] Relative student position in institutions matters. Here, then, the HBCUs can continue to have an important role in nurturing academic talent among black students.

Despite widely ranging methods, data sources, and professional disciplines, theories, and analyses, both studies that Commissioners contributed to the record after the briefing support a common conclusion. In short, HBCUs make a distinctive contribution to the well-being of black Americans.

# SUMMARY OF THE PROCEEDINGS

## Louis W. Sullivan

### *Historically Black Colleges and Universities and the Development of Students*

Dr. Sullivan stated that a majority of the nation's historically black colleges and universities (HBCUs) came into existence after the Emancipation Proclamation in 1863. Religious organizations built many of them with states taking over this task in later years.[47] Passage of the Voting Rights Act of 1965 and other momentous civil rights legislation in the

1960s, designed to prohibit discriminatory conduct and "eliminate the vestiges of segregation and discrimination," guaranteed equal rights in principle for African-Americans and other minorities.[48] There was, however, an unintended consequence: persistent questions arose in some quarters on the continued need for HBCUs and their educational effectiveness.[49]

The college years, Dr. Sullivan explained, are significant in the academic, social, and personal development of undergraduates. During this period, they make the transition from the certain, intimate world of the home to the more indeterminate, impersonal environment of the campus.[50] For some young African-Americans, he said, the development that occurs at HBCUs may have a profound influence on their lives, including how they perform roles such as spouse, parent, professional, or responsible citizen.[51] For example, African-American graduates of certain HBCUs successfully gained entry into graduate, medical, engineering, and law schools and other fields in percentages and numbers equaling or exceeding those of African-American students that attended wealthier, established predominantly white colleges and universities. Dr. Sullivan noted that HBCUs with such eminent track records include, for example, Xavier University, Spelman College, Morehouse College, Florida A & M, North Carolina A & T University, and Jackson State University.[52]

### *The Morehouse School of Medicine*

Dr. Sullivan explained that the Morehouse School of Medicine is a predominantly African-American medical school. Morehouse College established its medical school in 1975 specifically to increase the number of African-American and other minority physicians in Georgia and, more generally, the nation. The Morehouse School of Medicine, with modest financial resources, began as a two-year institution enrolling the first class of medical students in 1978. It expanded to a four-year school by 1981, and earned full accreditation in 1985.[53]

The institution, according to Dr. Sullivan, now counts 800 medical alumni, including a commissioner for health for the State of Georgia, a vice president of a large prestigious medical school, the personal physician of the President of South Africa, and physicians practicing in medically underserved rural and inner city communities. Dr. Sullivan noted that the institution's graduates pass national medical examinations at rates equaling or surpassing the overall national rate. Yet the Morehouse School of Medicine is one of the most recently established medical schools in the United States.[54]

### *Factors Accounting for the Educational Success of Students at Historically Black Colleges and Universities*

Dr. Sullivan identified four factors that help explain HBCU students' successful experience. These are:

1) Faculty members' dedication to teaching;
2) HBCUs' supportive social environment;
3) Strong encouragement to students to explore career options and leadership roles in their chosen careers; and
4) Faculty role models.[55]

Ending on a sober note, Dr. Sullivan said the vestiges of segregation and discrimination still linger in contemporary American society. Their elimination demands our sustained and dedicated efforts. The nation's HBCUs are significant in this regard, contributing to the academic, social, and personal development of many African-American and other minority citizens.[56]

## Earl S. Richardson

***Challenging the Rightful Place of Historically Black Colleges and Universities in Higher Education***

Dr. Richardson stated that the Civil Rights Act of 1964 (CRA) changed the landscape for HBCUs. According to him, following its enactment and federal enforcement in the early 1970s, discussions arose about the future role of HBCUs in contemporary higher education. Previously, the understanding of what HBCUs represented was clear: they were open to all students, and, according to him, were the only type of institution in the Southern states available to African-American students seeking a college education.[57]

Dr. Richardson said that during this historic period, an erroneous assumption that HBCUs restricted enrollment to African-Americans precipitated a discussion of HBCUs' rightful place in contemporary American higher education. In fact, he said, a high percentage of prominent African-American leaders in public life and the professions earned their college degrees from HBCUs because of the concentration of blacks in the South.[58] The State of Maryland illustrates this unique history. For example, Morgan State University counts among its graduates the "first black judge in Maryland, the first black congressman, the first [black] state senator, and the first [black] official to hold any statewide office." Dr. Richardson added that these leaders would not be in their positions without their HBCU attendance.[59]

***Impact of the Civil Rights Act of 1964 on Historically Black Colleges and Universities***

According to Dr. Richardson, the CRA required traditionally white institutions to increase minority enrollment. Traditionally white institutions sought to meet this requirement through financial incentives to mostly poor black students, who were thus able to enroll in such institutions for the first time.[60] He noted that there was no parallel requirement for HBCUs to attract white students. He suggested that the federal government missed an opportunity to create a fully desegregated higher education system by adopting this approach to integration.[61]

At the same time, Dr. Richardson said, HBCUs were not on par with traditionally white institutions in many important respects, such as the quality of facilities and range of academic programs.[62] He opined that Southern states deliberately neglected HBCUs as part of a strategy to limit educational opportunities for blacks at all levels.[63] He estimated that by the time the federal government enforced CRA, most HBCUs were about a century behind traditionally white institutions.[64] HBCUs therefore were not competitive in attracting their traditional constituency and even less so with respect to white students.[65] In the early 1970s, Dr. Richardson noted that black enrollment at HBCUs declined substantially as did the smaller white enrollment.[66] With regard to Morgan State University, he said the graduate student population was once slightly majority white, but became almost entirely black within a decade. He further noted that the institution's enrollment only recently re-attained its previous heights.[67]

***How the Historically Black Colleges and Universities Survived***

Dr. Richardson suggested that four factors contributed to the survival of HBCUs:

1) In the mid-1970s, in light of the devastation wrought on HBCUs' enrollment, the federal government required states to enhance such institutions as a precondition to developing desegregation plans.[68] These plans were intended to "create parity and comparability between ... black ... [and] white institutions so that they could be equally competitive to students regardless of ... [the] race [of the student body]," a task that Dr. Richardson said is still incomplete;[69]
2) There is a core group of African-American students who choose to attend HBCUs over traditionally white institutions;[70]
3) Growth in the African-American population since the 1960s translated into a growing number of college-age students by the mid-1980s. On the other hand, the white college-age population had been declining, and would continue to do so for another decade;[71]
4) Traditionally white institutions became more selective in their admission requirements. African-American students, on average, were less competitive on standardized tests and other measures of pre-college academic achievement than their white counterparts were. This compelled more and more African-American students to seek admission in institutions with liberal admission policies, including HBCUs and community colleges, which led to substantial growth in the enrollments of such schools.[72]

***Historically Black Colleges and Universities are Primary in Narrowing the Higher Educational Gap between African-Americans and Whites***

Dr. Richardson stated that African-American educational attainment lags far behind that of whites despite HBCUs' success in enabling college access. In 1970, a year that marked the beginning of a federal emphasis on "parity, equity, and affirmative action," he said the difference between whites and African-Americans holding at least a bachelor's degree among persons 25 to 29 years of age favored whites by 10 percent. [73] At that time, attaining parity necessitated increasing the number of African-Americans with at least a bachelor's degree by 160,000. [74] Dr. Richardson pointed out that the trend in recent years has been an ever-widening educational gap between the two groups.[75] A similar comparison made today shows a 17-point difference between the groups because whites exhibit a higher level of degree attainment.[76] Attaining parity now requires approximately 400,000 additional African-Americans with bachelor's degrees, far more than the necessary number in 1970.[77]

Dr. Richardson stated that less selective or open-access campuses such as HBCUs would have to shoulder the enormous task of increasing the numbers of minority students enrolling in and graduating from college if this goal were to be met.[78] He reasoned that:

1) With growing emphasis on selectivity, traditionally white institutions are increasingly recruiting students that scored in the upper ranges of the SAT.[79] Less than 1 percent of blacks as compared to 6 percent of whites score 1400 or higher on the SAT, and less than 5 percent of blacks as compared to 24 percent of whites score 1200 or higher. HBCUs and less selective institutions therefore represent "realistic" options when it comes to enrolling minority students;[80]

2) Selective campuses generally graduate a percentage of African-Americans that is half that of the total population from which they recruit. For example, the Ivy League institutions, all of which are national universities, collectively have a graduating class that is 7 percent African-American; the graduating class of the University of California, Berkeley and the University of California, Los Angeles combined, 4 percent; and the graduating class of the University of North Carolina, Chapel Hill, 11 percent. These percentages constitute about half the proportion of blacks in the nation, California, and North Carolina, respectively. Thus, the nation cannot look to selective campuses to increase significantly the number and percentage of black and other minority graduates;[81]
3) HBCUs, in contrast, produce a larger number of black graduates than selective institutions. Twenty-four individual HBCUs graduate the same numbers of blacks as the University of Michigan and the University of North Carolina, Chapel Hill combined. Morgan State University alone graduates almost as many black students as all of the Ivy League campuses taken together, and Southern University graduates even more. Howard University and Florida A & M graduate more blacks than the entire University of California System, the largest public university system in the nation. "HBCUs [therefore] must be an integral part of any strategy for increasing the number of black graduates in the nation."[82]

In light of the reasons he presented, Dr. Richardson was emphatic that allocation of additional resources to HBCUs would have the greatest impact in narrowing the degree attainment gap between minority and majority students.[83]

## Jamie P. Merisotis

### *Four Areas of HBCU Effectiveness*

According to Mr. Merisotis, any review of the educational effectiveness of HBCUs should consider four areas. They are:

1) HBCUs as leaders in student engagement;
2) HBCUs as community-based institutions that "promote civic engagement and service learning;"
3) HBCUs as drivers of educational attainment for "low income, first generation, and disabled students;" and
4) HBCUs as examples of success in a national effort to "improve the quality of science, technology, engineering, and mathematics" (STEM) in education and research.[84]

### *Serving As Leaders in Student Engagement*

Mr. Merisotis stated that a 2004–2005 National Survey of Student Engagement (NSSE) found "more supportive learning environments for students" on HBCU campuses.[85] For example, compared to African-American students enrolled in "predominantly white institutions," those attending HBCUs reported more frequent interactions with faculty members.[86] They also reported greater likelihood of voting and believed that their institutions promoted "personal spiritual growth."[87]According to Mr. Merisotis, the NSSE results in conjunction with other relevant research support the claim that HBCUs offer an educational experience that encourages a high level of student engagement and "enhances [students'] intellectual gains and accomplishments."[88]

***Promoting Civic Engagement and Service Learning***

Mr. Merisotis noted that the extent to which colleges and universities contribute to student "civic engagement and participation in ... democratic institutions, such as voting, volunteering, and community involvement"[89] is of general interest. Institutions may affect this in a number of ways, one of which is to require service learning.[90] The 2004 Campus Compact[91] membership survey, according to Mr. Merisotis, documented that HBCUs and other minority-serving institutions are more likely to have service and service-learning requirements as a condition of graduation, community service or service learning offices, directors of community service or service learning, and partnerships with kindergarten through secondary schools and faith-based organizations.[92] He said that such deliberate cultivation of civic engagement among HBCU students "point[s] to a profoundly different approach to student success." He noted that the national literature on service learning and civic engagement has yet to document adequately the effects of HBCUs' intentional strategies.[93]

***Enabling Educational Attainment for Low Income, First Generation, and Disabled Students***

HBCUs, Mr. Merisotis said, offer disabled students and those from educationally and economically disadvantaged backgrounds "additional support, guidance, and mentoring" to ensure admission into and graduation from college.[94] According to him, the Council for Opportunity in Education reported that almost 75 percent of HBCUs participate in the TRIO[95] program compared to less than 25 percent for non-HBCUs. The TRIO programs on these HBCU campuses collectively serve about 70,000 students and receive approximately $70 million in federal funds.[96]

***Science, Technology, Engineering, and Mathematics (STEM)***

Mr. Merisotis stated that study of science, technology, engineering, and mathematics (STEM) plays a critical role in enhancing America's global and economic competitiveness. He offered that African-Americans are seriously underrepresented as a percentage of the STEM workforce and as a percentage of students enrolling in and graduating from college-level STEM programs. In his view, enhancing the quality and success of HBCU STEM programs is an appropriate way to improve the nation's strength in these fields.[97]

Mr. Merisotis reported that the undergraduate level initiative known as "HBCU UP" enhances the success of HBCU STEM programs. The National Science Foundation (NSF) funds the program at 14 HBCU sites and supports a wide range of activities, including "curriculum enhancement, faculty professional development, undergraduate research [to enable acquisition of research experience], collaborations with research institutions, and other activities that meet institutional needs."[98] NSF, according to Mr. Merisotis, reported that institutions with "HBCU UP" programs in place for five years have improved passing rates in mathematics courses essential to success in STEM programs, such as Algebra, Pre-calculus, and Calculus I.[99] Performance in other required STEM courses, such as Biology I and Physics I, also showed improvement. Of the STEM graduates from these HBCUs, about 25 percent have acquired undergraduate research experience that will help them in graduate school.[100]

Mr. Merisotis said that the four factors discussed above illustrate the multidimensional nature of educational effectiveness. A fair and comprehensive assessment of HBCU performance should extend beyond examinations of rates of graduation, retention, or job placements, which are necessary indicators but not fully representative of the benefits of attending such institutions. He further pointed out that a multidimensional approach would "improve the targeting of strategies to continuously upgrade quality and performance at these nationally essential institutions of higher learning."[101]

## Raymond C. Pierce

### *A Real and Continuing Threat to Historically Black Colleges and Universities*

Dean Pierce stated that HBCUs continue to effectively educate large numbers of African-Americans, thus contributing to the nation's need for a learned population and skillful workforce. On the other hand, he said there is no evidence that their closure would result in extant students enrolling in and graduating from traditionally white institutions.[102] He also declared that there is a "real and continuing threat" to HBCUs in part because the federal government refused to enforce federal civil rights laws that pertain to African-Americans enrolled in public HBCUs.[103] He further pointed out that because of HBCUs' critical educational contribution, and because full participation in the world economy demands an educated workforce, this threat directly undermines the nation's global competitiveness.[104]

### *Desegregation in Higher Education*

Dean Pierce stated that federal civil rights efforts involving HBCUs occurred after the 1954 desegregation litigation in *Brown v. Board ofEducation*.[105] He pointed out that desegregation in state higher education systems involved private lawsuits seeking equal funding for HBCUs relative to traditionally white institutions in combination with federal government intervention.[106] The Adams cases of the 1970s[107] are illustrative. Private individuals filed suit in federal court against the U. S. Department for Health, Education, and Welfare (HEW) for failure to enforce federal civil rights laws with regard to the equality of African-Americans enrolled in HBCUs.[108] According to Dean Pierce, HEW's Office for Civil Rights (OCR) had determined that 19 Southern and border states violated Title VI of the Civil Rights Act of 1964 for "failure to equally protect the rights of African-Americans attending [HBCUs] pursuant to the Fourteenth Amendment of the United States Constitution."[109] He noted that these states nonetheless continued to treat HBCUs unequally *vis-à-vis* traditionally white institutions in a number of areas, including resource allocation.[110] Furthermore, federal funds flowed uninterrupted to these states, thereby allowing them to operate segregated institutions of higher education.[111] In the Adams cases, Dean Pierce said, the court found against OCR and directed it to develop guidelines to desegregate state systems of higher education. OCR required the 19 Southern and border states to submit plans that would bring about compliance with Title VI.[112]

According to Dean Pierce, 14 of the 19 states presented plans acceptable to OCR and entered into agreements to implement five-year Title VI compliance plans.[113] Of the remaining five, *four* of them—Louisiana, Tennessee, Alabama, and Mississippi—failed to reached agreement with OCR, and each went on to litigate in federal court. Dean Pierce stated that Mississippi's case advanced to the U. S. Supreme Court as *Ayers v. Fordice* in 1992. OCR referred Ohio, the *fifth* state, to the U.S. Department of Justice (DOJ) for litigation, when the parties could not agree on a plan to desegregate Ohio's system of higher education. DOJ, however, did not file the case in court.[114]

### *The 1978 Civil Rights Policy for Higher Education Desegregation*

Dean Pierce explained that OCR's 1978 civil rights policy, "Revised Criteria for the Desegregation of State Systems of Higher Education," used whether or not states were enhancing their HBCUs as one factor in determining state compliance with desegregation requirements. He said that HBCUs' limited academic program offerings and substandard physical facilities were typical at that time and resulted largely from the circumstances of their founding and existence. Enhancement consisted of establishing and locating attractive academic programs on HBCUs' campuses to expand educational opportunities for African-American students and attract a diverse student body. The policy's second component for determining states' compliance was whether or not traditionally white institutions used affirmative action in recruitment and scholarship to attract and enroll more African-American students. The intent of the 1978 policy, according to Dean Pierce, was to make academic offerings, rather than the racial composition of the student body, the primary basis for students' choice of HBCUs. If successful, the 1978 policy would have created a unitary system of higher education in place of racially identifiable black and white institutions.[115]

Dean Pierce stated that in 1988 Secretary William Bennett of the U. S. Department of Education directed its Office for Civil Rights (OCR) to review the 14 states that had entered into agreements in 1978 to ascertain progress in compliance with Title VI of the 1964 Civil Rights Act. OCR employed a checklist analysis[116] to make that determination and concluded that eight of the 14 states were indeed in compliance.[117] According to Dean Pierce, OCR did not conduct compliance reviews for the remaining six states.[118]

### *The 1994 Federal Policy for Higher Education Desegregation*

Dean Pierce stated that the U. S. Supreme Court ruled in *United States v. Fordice* that "states have an affirmative duty to remove all vestiges of the past practice of segregation that have a present day effect to the greatest extent practicable."[119] In response to the Supreme Court decision, Mississippi initially suggested desegregating its higher education system by closing all of the state's HBCUs.[120]

Dean Pierce said that in 1994, following *Fordice*, OCR created a far more rigorous policy for higher education desegregation than its 1978 version. Dean Pierce explained that under the 1994 policy, OCR employed a vestige analysis to determine if states previously found in violation of Title VI were currently in compliance. Specifically, it sought to ascertain if states had to the "greatest extent practicable addressed and removed all of the vestiges that have a present day effect on the educational opportunities of African-Americans attending historically black colleges and universities."[121] Subsequent to publishing its 1994 policy, OCR obtained newly signed five-year compliance agreements from seven states—Ohio and the remaining six of the 14 states that entered into agreements with OCR under the 1978 policy.[122] Notably, Dean Pierce said, the 1994 agreements did not include traditionally white institution affirmative action programs to attract African-Americans, focusing instead on enhancing HBCUs and addressing "the remaining vestiges of [segregated higher education] found to have continuing and present day effects."[123]

### *The Present*

In Dean Pierce's view, the "problem we face today is almost a revisitation of the situation that led to the Adams cases in the 1970s."[124] In 2000, 2001, and especially 2002, there were indications that the seven states that signed the new agreements were backing away from them. Ohio in particular was seeking to revert to policies that would negatively affect Central State University, a historically black university, potentially in violation of Title VI. Dean Pierce suggested that the State of Maryland faces a similar situation. These states, though, continued to receive federal funds for public higher education.[125]

At the same time, Dean Pierce said strong evidence exists that the eight states found in compliance with Title VI in 1988 are now engaged in actions that negatively affect some HBCUs, such as unnecessary program duplication.[126] Historically, program duplication served to create and maintain a segregated higher education system. Dean Pierce predicted: "[L]itigation that was brought in the Adams cases will once again find its way to the courts."[127]

## Mikyong Minsun Kim

### *Four Areas of Focus*

Dr. Kim stated that she studied the educational effectiveness of HBCUs. To do so, she employed longitudinal, national-level data sets from the Higher Education Research Institute (HERI), which is located at the University of California, Los Angeles.[128] She concentrated on four areas: (1) comparisons of institutional and student characteristics between HBCUs and historically white colleges and universities, (2) review and comparison of her research findings with those of other studies, (3) discussion of HBCUs' contribution to the development of African-American students, and (4) evaluation of whether HBCUs merit strong support.[129]

### *Institutional and Student Characteristics of Historically Black Colleges and Universities and Historically White Colleges and Universities*

Dr. Kim indicated that the American higher education community includes 103 HBCUs of which 89 are four-year institutions. Of these 89 institutions, 41 are public and 48 private. Collectively, the 89 HBCUs award about 30 percent of the bachelor's degrees that African-Americans earned. Dr. Kim's research also showed that a high percentage of African-American political leaders, professionals such as lawyers and doctors, and doctoral degree recipients are graduates of HBCUs.[130]

Dr. Kim reported that HBCUs tend to have fewer resources compared to historically white colleges and universities, such as lower expenditures for each full time equivalent student, lower average faculty salaries, and poorer physical facilities.[131] HBCUs also tend to have smaller enrollments, a lower student-faculty ratio, and higher student-faculty interactions, characteristics that other panelists noted as well.[132] She pointed out that research has shown that these same institutional characteristics tend to be positively associated with student development.[133]

Continuing, Dr. Kim stated that students enrolled in HBCUs were generally less academically prepared than historically white college and university students.[134] Other research, such as that of Elliott, et al., reported a similar but more detailed observation: "accounts of race, ethnicity, and science made it clear that [African-Americans] are relatively low on most measures of preparation and developed ability, and that these deficits begin early in their school careers. They are concentrated just before the point of entrance to college. Both the average SAT mathematics ... scores and the math and science proficiencies of the twelfth-grade blacks are about a standard deviation [a statistical measure of the breadth of African-Americans' range of scores] behind ... those of whites. Thus, black grade-12 achievement in math is about the same as, and in science a little worse than, white grade-8 achievement."[135] In their study on the causes and consequences of attending HBCUs, Fryer and Greenstone found that blacks enrolled in traditionally white institutions generally have substantially stronger academic credentials than those attending HBCUs.[136]

Dr. Kim stated that the family backgrounds of students enrolled in HBCUs tended to be less affluent. She noted that this finding is consistent with other researchers' reports.[137] In addition, she said African-American students at HBCUs were close to 1.5 times more likely than African-American students enrolled in historically white colleges and universities to participate in faculty members' research. Dr. Kim said such engagement is a predictor of degree completion.[138] Based on HERI data, she reported that the African-American degree completion rate at HBCUs was 55 percent compared to 63 percent at historically white colleges and universities,[139] although the college grade point averages of African-American students in both types of institutions did not differ.[140]

***Research Findings on Selected Outcomes***

Dr. Kim examined the impact of students' attending HBCUs and historically white colleges and universities on three academic outcomes: (a) their overall academic ability, (b) writing ability, and (c) mathematics ability.[141] Specifically, she examined a sample of African-American students who earned bachelor's degrees from either HBCUs or historically white colleges and universities with data that included information covering a nine-year period after their graduation.[142] Dr. Kim studied institutional effectiveness using sophisticated statistical procedures.[143]

Dr. Kim found no significant difference between African-American students in HBCUs and historically white colleges and universities in the three outcome areas.[144] These findings are somewhat consistent with research literature produced more than a decade ago.[145] Similarly, she found no significant difference in the likelihood of earning a baccalaureate degree between African-American students in HBCUs and historically white colleges and universities. In this respect, she said her finding contradicted those that several scholars reported more than 10 years ago.[146] Her finding that African-American students in HBCUs and historically white colleges and universities faced no significant difference in early career earnings echoed the results of some previous studies but contradicted others.[147] Dr. Kim viewed her findings of no significant differences in a positive light because attendance at HBCUs or historically white colleges and universities has no effect on African-American students' academic success and early career earnings.[148]

With respect to earnings, Fryer and Greenstone's 2007 study may be of interest. The authors reported that "in the 1970s, HBCU matriculation [on the part of African-Americans] was associated with higher wages, relative to attending a traditionally white institution. By the 1990s, however, there was a substantial penalty. In fact, there is a statistically significant 20 percent decline in the relative wages of HBCU graduates between the two decades. [The data] provide some support for the possibility that HBCUs' relative decline is partially due to improvement in traditionally white institutions efficacy in educating blacks, but this evidence certainly [is not] decisive. This question of why HBCUs' performance declined merits further research."[149] One should view these findings with the authors' caveat in mind: "The results [here referring to the sum of their findings] are robust across ... [the] four [statistical] approaches [employed in the study]. However, lacking a randomized experiment or credible quasi-experiment, thorny issues of selection [of cases for study] may remain. Consequently, we urge caution in interpreting the results as causal."[150]

### *Historically Black Colleges and Universities and Student Engagement and Development*

Dr. Kim observed that African-American students in HBCUs were more deeply engaged in the academic community than their counterparts at historically white colleges and universities. She also pointed out, as did other panelists, that the environment at HBCUs was strongly supportive of students' academic growth.[151] Historically white colleges and universities, on the other hand, are able to offer more resources to their students.[152] Thus, she posited that HBCUs and historically white colleges and universities contribute to student learning in different ways.[153]

### *Historically Black Colleges and Universities Merit Support*

In considering whether HBCUs have a significant impact on academic outcomes and early career earnings given students' weaker pre-admission academic profiles and the institutions' severely constrained resources, Dr. Kim believed that the issue merited further investigation.[154] She concluded that the "findings of [her] studies and of other reports lend support to the proposition that HBCUs contribute significantly to higher education in this country and merit strong support."[155]

## Discussion Summary

### *Academic Outcomes at Historically Black Colleges and Universities*

Commissioner Kirsanow noted that despite HBCU students' generally lower SAT scores, Dr. Kim found no significant differences between HBCUs and historically white colleges and universities in developing the writing, mathematics, and general academic abilities of students.[156] He asked Dr. Kim if her study controlled for the nature of an institution (for example, whether an institution is urban, suburban, or rural), because a significant difference might occur between HBCUs and historically white colleges and universities in urban settings. Dr. Kim stated that she did not control for regional effects, but controlled for institutional selectivity, size of enrollment, SAT scores, high school grade point average, pre-test measures, and parental education and income levels.[157] The lack of a control for regional effects notwithstanding, Commissioner Kirsanow argued that, HBCUs were "doing a much better job than traditionally white universities in educating African-American students"—more than the findings of no significant difference suggest—given the institutions' generally lower funding levels and the academic under-preparedness of their students at enrollment.[158]

Dr. Richardson agreed with Commissioner Kirsanow's assessment, but stressed the importance of comparing similar types of institutions. Morgan State University's graduation rate at the end of six years is 43 percent and seemingly modest. Yet when matched against equivalent urban universities, he stated that Morgan State University's retention and graduation rates exceeded theirs.[159] Dean Pierce similarly maintained that institutional comparability remains important in the assessments. For example, he noted that North Carolina Central University School of Law is different from Howard University School of Law and equivalent to the Southern University School of Law, Texas Southern University Thurgood Marshall School of Law, and Florida A & M University.[160]

### *Dropout, Graduation, and Performance on National Examinations*

Commissioner Kirsanow raised the concern of dropouts from the nation's five black law schools. He noted that 43 percent of enrolled black law students leave without earning a degree.[161] Dean Pierce said that North Carolina Central University School of Law has a higher attrition rate particularly among African-American males. He explained that the North Carolina Central School of Law accepts students with lower Law School Admissions Test (LSAT) scores who nonetheless have high college grade point averages. Notably, the institution's law school also enrolls large numbers of students.[162] Likely contributing to the higher attrition rate, Commissioner Kirsanow postulated, is the greater need for financial aid among HBCU students. Some 84 percent of HBCU students compared to 55 percent of white students at traditionally white institutions require financial aid to attend college.[163]

Dr. Sullivan stated that Spelman College's graduation rate of 77 percent exceeds, for example, those of Bates College, Colby College, University of California at Berkeley, University of California at Los Angeles, University of Michigan, Claremont College, and Carnegie-Mellon University—all of which are [predominantly] white institutions with greater financial resources. He also pointed out that students at the Morehouse School of Medicine performed better on the U. S. Medical Licensing Examination than all other medical students in the nation despite lower SAT scores on entry. Part of the reason, according to Dr. Sullivan, is recognition that student potential may not have been developed, frequently because of the high schools attended.[164]

Commenting on a different national-level examination, Dean Pierce stated that in 2005, the first-time bar passage rate of North Carolina Central University School of Law's graduates was 81 percent, tying with Duke University School of Law. The law school produced this achievement despite fewer resources and enrolling students whose pre-entry academic profiles predicted lower achievement.[165]

### *Historically Black Colleges and Universities as Pipeline to Graduate School*

Vice Chair Thernstrom expressed admiration for HBCUs. Several years ago, she visited Savannah State University and found that many of its African-American students had limited academic skills on entry, a reflection of the "racial gap in academic achievement at the end of high school."[166] Savannah State University's dedicated efforts to improve students' academic abilities impressed Vice Chair Thernstrom, particularly the quality of education that the institution offered.[167]

Vice Chair Thernstrom recalled that perhaps nine of the 10 top institutions sending African-Americans to graduate education were HBCUs. Commissioner Kirsanow said that the top 20 institutions in sending African-Americans to graduate school were HBCUs. He added that the figure varied by academic discipline. Among the STEM disciplines, such as biology, 12 of the 15 institutions sending the largest number of African-Americans to graduate school were HBCUs, while in the physical sciences, the comparable figure was 14 or 15 out of 15. Such statistics, Vice Chair Thernstrom said, affirmed HBCUs' quality of education.[168]

***Historically Black Colleges and Universities, STEM, and Teacher Education Programs***

Commissioner Kirsanow found it striking that HBCUs awarded 40 percent of all black graduate STEM degrees.[169] He noted that the top 30 HBCUs graduated the largest numbers of blacks that enrolled in STEM doctoral programs, despite variations by gender.[170] Panelists offered several suggestions for this phenomenon. Dr. Sullivan speculated that faculty commitment played a role. He pointed to Xavier University, which, despite marginal resources, sent more black graduates to medical schools than any other institution in the country, black or white. He referred to a long-time Xavier faculty member, Professor Carmichael, who was renowned for spending an "inordinate amount of time" with Xavier University students. Such faculty members are the reason for Xavier University's success, he suggested.[171] Mr. Merisotis said that HBCUs are open access institutions responsive to market and community needs. Thus, they channel students into STEM programs in which the labor force need is great.[172]

Dr. Richardson stated that the culture on HBCU campuses is a critical factor in their remarkable productivity in the STEM fields. Although both HBCUs and non-HBCUs provide academic support programs to assist minority students, he observed that predominantly white schools offer more of a collection of individual programs. HBCUs, on the other hand, integrate such programs into a nurturing culture that permeates the entire university. Dr. Richardson discerned that HBCUs send a message that the institution will help students every step of the way. This fosters a campus culture that reaffirms that dedicated students can and will succeed, regardless of their SAT scores.[173]

Role modeling, Dr. Richardson stated, also helps explain HBCUs' success in the STEM disciplines. Dr. Richardson said HBCUs convey to students that their professors and administrators come from similar economic backgrounds, and if the professors and administrators succeeded, so too can the students.[174] Dr. Kim also emphasized the importance of role modeling. She observed that on HBCU campuses, African-American teaching assistants and professors in engineering, mathematics, and science serve as "special role models" to students. Of interest here is Dr. Kim's research finding that undergraduate African-American students on HBCU campuses are approximately 1.5 times more likely to be involved in faculty members' research than their counterparts in historically white colleges and universities.[175]

Mr. Merisotis shifted the discussion to HBCUs' role in teacher education. Almost half of the African-American teachers in today's schools trained in such institutions, but this accomplishment has not been sufficiently acknowledged, he said.[176] The need for African-American and minority teachers is greater than ever, given current and predicted future demographic trends. Mr. Merisotis called for investment in HBCUs to ensure their success in educating African-American teachers.[177]

***Historically Black Colleges and Universities As Catalyst for Change***

According to Dr. Richardson, HBCUs not only prepare students for graduation but also serve as catalysts for systemic change in higher education. In 1980, for example, Maryland had graduate engineering programs located at Johns Hopkins University, the United States Naval Academy, and the University of Maryland, all of which are non-HBCUs. Yet, he noted that in 1981 about 20—less than 1 percent—of the engineers in the state were African-American. Today, the comparable figures are about 150 or 19 percent. Dr. Richardson attributed the increase to the state establishing an engineering program at Morgan State University in 1984. Notably, Morgan's program graduated about 100 of the 150 African-American engineers.[178]

Dr. Richardson pointed out that historically, scholars explained the dearth of African-American engineers by the paucity of African-Americans who could successfully complete an engineering program. According to him, by its action and achievement, Morgan State University compelled non-HBCUs to "find ... blacks and [enroll] them in [their] engineering schools."[179]

***A Nurturing Environment***

Chairman Reynolds commented that HBCUs as well as some TWIs enroll students who lack an academically rigorous education, yet traditional white schools seem to have a much higher attrition rate.[180] He postulated that these institutions might use different teaching models. Chairman Reynolds asked whether "it could make a significant difference if a student, especially a student who has not received a rigorous preparation, [enrolls] in a college that focuses on research [and makes heavy use of] teaching [assistants] as opposed to [one in which] students are [principally taught by professors]."[181]

Dean Pierce stated that HBCUs, both historically and presently, provide a nurturing environment in which a culture dedicated to engaging students within and outside the classroom flourishes. He said such an environment enabled North Carolina Central University School of Law to match Duke University School of Law's bar passage rate in 2005, despite its students' lower LSAT scores. Consequently, Dean Pierce deliberately encouraged his faculty members to spend more time engaging with students to improve student performance than on producing scholarly articles for law journals.[182]

The Elliott, et al., article reported that"... persistence is not just a matter of average preparation, but of competitive position as well: a reasonably well-prepared student at an HBCU ... would be in a strong competitive position in his or her institution[al environment] ..." and thus more likely to succeed.[183] The article concluded that HBCUs have a stronger record of producing science graduates at the undergraduate and doctoral levels than more elite, predominantly white institutions. This success occurred despite evidence that black students enrolling in HBCUs are, on average, less prepared than those entering elite institutions.[184]

Vice Chair Thernstrom asked Dr. Sullivan if any curriculum distinction existed between Morehouse School of Medicine and other medical schools. Dr. Sullivan responded that the critical factor is not curricular, but rather the medical school faculty members' commitment to student success. One indication of this is the enormous amount of time faculty members spend with students. Another factor, Dr. Sullivan stated, is an environment that encourages students to "take risks, [including the] risk [of] asking question[s]" without fear of embarrassing themselves. The learning process demands not only a good teacher, but also a student who is willing to engage. He said successful HBCUs create an environment that encourages student participation. Yet another factor, one that made the Morehouse School of Medicine distinctive, is encouraging students to think of themselves as leaders. Dr. Sullivan stated that all of these factors are found at other HBCUs.[185]

Chairman Reynolds suggested that perhaps HBCUs teach at the pace of their students. He observed that the basis for this suggestion was that at schools that are traditionally white, the white students' generally higher level of academic preparedness permitted faculty members to teach at a faster pace, which had a negative consequence on the black students who are less academically prepared. He suggested the pace of teaching at HBCUs was more in sync with black students' level of academic preparation.[186] Dr. Sullivan agreed, noting that he would have to modify his earlier statement of no curricular difference between the Morehouse School of Medicine and other medical schools. First year medical students arrive on campus in mid-July, the earliest opening date of any medical school. He said this gave the institution time to raise the level of academic preparedness of students with deficiencies. The medical curriculum itself begins in September and is similar to those of other medical schools. Dr. Sullivan added that faculty members often help students that experience academic difficulty, highlighting again the criticality of faculty commitment.[187]

### *Lack of Parity between HBCUs and Non-HBCUs*

Vice Chair Thernstrom asked Dr. Richardson to clarify what he meant when he spoke of "... creating a parity between black and white institutions." Dr. Richardson explained that parity involved provision of "equitable resources ... [that are] consistent with the magnitude of the task." He stated that the average education cost per student is higher at HBCUs than at white institutions, giving rise to the perception that HBCUs have more resources than white institutions. Dr. Richardson was emphatic that this was manifestly incorrect. The task of educating academically under-prepared students imposes a heavier workload on faculty members and staff from the moment of their admission.[188] He offered four examples to debunk the notion of more resources at HBCUs.

### *Subsidizing Student Tuition*

First, Dr. Richardson stated that African-American students enrolled in HBCUs almost always need extensive financial assistance. The only source of support for many of them is the federal Pell Grant that today supports less than half the total cost of college attendance. Dr. Richardson said that some HBCUs use monies collected from tuition and fees to make up the difference so that students are able to enroll. The institution would ordinarily use these funds for operating costs, he pointed out. Subsidizing tuition is indispensable for student retention, but reduces the funds available to operate the institution. This seriously undermines necessities, such as hiring more faculty members and purchasing new equipment.[189]

### *Greater Staff Workload*

Second, Dr. Richardson argued that staff workload at HBCUs is greater than that at historically white campuses of similar size and levels of staffing and resources. Because of their limited financial resources, many students who attend HBCUs lack access to the Internet to apply, register, and pay tuition and fees. Thus, students complete each of these processes manually and generate a "horrendous" need for staff time. The situation is exacerbated when staff must repeat the payment process monthly because almost all HBCU students participate in installment payment plans. Such time-intensive payment methods, Dr. Richardson said, are less common at comparable majority institutions. Their students are more likely to come from middle and upper middle class families and hence have the financial capability to use the Internet for application, registration, and payment of tuition and fees. Majority institutions' need for staff time to manage such processes is consequently less. The nature of disparity between HBCUs and majority institutions is not always readily apparent.[190]

***Lack of Adequate Facilities***

Third, Dr. Richardson stated that facilities at HBCUs are not adequate. The State of Maryland has two state-supported architectural programs, one at a traditionally white institution, housed in a state-of-the-art building, and the other at Morgan State University. The 30-year old architectural program at Morgan State University does not have its own facility and, according to Dr. Richardson, was moved to different buildings on campus to ensure its accreditation. He noted that this disparity denies architectural students at Morgan State University an equal opportunity to experience state-of-the-art facilities.[191]

***Lack of Technology Transfer and Commercialization Centers***

Fourth, Dr. Richardson pointed out that in the State of Maryland, engineering and business programs have well-established technology transfer and commercialization centers. Two such centers are on the majority institution campuses with engineering and business programs. Dr. Richardson stated that Morgan State University's engineering program, despite being the most recently established, graduates the largest number of baccalaureate and doctoral African-American engineers in the state. Morgan State University also has the proper combination of engineering and business programs, but despite the institution's request, the state has not provided faculty, facilities, or other support comparable to what it makes available to the other two public campuses with similar engineering/business program combinations.[192]

***Faculty Salaries***

Vice Chair Thernstrom raised the question of HBCU faculty salaries. Dr. Richardson said that faculty salaries are dependent on the amount of resources available. In the main, HBCU faculty salaries still lag behind those of majority institutions, even when the comparison is within a similar institutional classification.[193] Morgan State University's faculty salaries fit this pattern, but they compare favorably with other HBCUs. Dr. Richardson stated that most of Morgan State University's faculty members have doctoral degrees from the same prestigious universities as faculty members at any other school, but limited financial resources prohibit the institution from offering competitive salaries. This is a significant obstacle to the institution as it competes for the best faculty minds with better-endowed schools in similar institutional classifications.[194]

***The Right to Exist***

Vice Chair Thernstrom noted that academic scholarship on elementary and secondary education still calls for integration, including the use of busing to achieve this goal. She suggested that proponents of such views would have serious reservations about the continued existence of HBCUs.[195] She asked Dean Pierce if any "politically significant voices" still challenge the need for racially identifiable colleges and universities.[196]

Dean Pierce replied that he discounted such voices. Segregation (i.e., racially identifiable schools) at the kindergarten through secondary school levels is distinct from racially identifiable institutions in higher education because students in the former are assigned to schools by district, whereas college attendance is voluntary. However, the resources available at colleges competing for the best applicants influence students' decisions about college attendance.[197]

Vice Chair Thernstrom further asked if the civil rights community, particularly among black spokespersons, was split on integration when the Supreme Court decided *Fordice* in 1992.[198] Dean Pierce responded that many within the civil rights community expressed some concern that the Supreme Court decision in *Fordice* might result in states placing the burden of higher education desegregation on HBCUs.[199]

Dean Pierce stated that Supreme Court Justice Clarence Thomas' concurring opinion in *Fordice* illustrated this tension. On the one hand, Justice Thomas opposed the creation of HBCUs as "enclaves" for the black community. At the same time, he pointed out how unfair it is for HBCUs, which bore the burden of segregation in the past, to again shoulder responsibility for present day desegregation efforts. Dean Pierce remarked that some voices continue to argue that removal of the vestiges of past segregation requires merging or shutting down HBCUs. Fortunately, federal policy now directs OCR to "strictly scrutinize any state effort to close or merge an historical black college or university in the desegregation process ... as long as [there are] outstanding Title VI violations." Dean Pierce recognizes that the qualification is necessary and that state fiscal considerations might force the merger or closure of some public colleges in the future.[200]

***Struggle over State Funding and Program Duplication***

Remarking on Vice Chair Thernstrom's earlier reference to politically significant voices "arguing against HBCUs," Dean Pierce claimed to have heard such opinions in the state assemblies and legislatures.[201] They are not arguing against the existence of HBCUs *per se,* because these institutions are segregated by choice rather than by law. According to him, the belief that HBCUs have funds that traditionally white institutions need underlies these voices. For example, "Ohio State University needs [the] money that Central State University has, or the University of Mississippi needs [the] money that Alcorn State has, or the University of Georgia needs [the] money that Savannah State has."[202] Higher education, Dean Pierce remarked, is a very costly enterprise and competition for limited state funds is aggressive.[203]

Vice Chair Thernstrom asked whether racially identifiable schools were an element in the public argument against HBCUs' existence. Dean Pierce assented but deemed such an argument as misleading.[204] He pointed out that students choose Savannah State University because the school offers a nurturing environment and a better chance of graduating, not because, for example, the University of Georgia barred the attendance of blacks, as it did in a different era.[205] He said that state legislators claim that Savannah State University is a segregated institution and should not be maintained as such. The real issue, he said, is the reduced funding for higher education. According to Dean Pierce, legislators resolve this situation by "back[ing] off agreement[s] to enhance or strengthen historically black colleges and universities so that more money [is available] for the traditionally white institutions." He submitted that state legislatures and assemblies duplicate programs to diminish the effectiveness of HBCUs and build up traditionally white institutions to further support the argument for closing black colleges.[206] This explains the growing program duplication that is contrary to established federal policy and case law.[207] He also mentioned that the State of Georgia is replicating an academic program that Savannah State University has in a geographically proximate public two-year community college, thus placing the university's unique attraction for good students "under a significant threat." In addition, program duplication now threatens Morgan State University, Bowie State University, Kentucky State University, and other institutions.[208]

Commissioner Taylor stated that while serving in the Attorney General's Office in the State of Virginia, he resolved a case involving a review of program duplication in state higher education institutions.[209] He discovered that when resources enabled the creation of excellent and highly desirable academic programs, states rarely placed them on an HBCU campus.[210] Commissioner Taylor also said that the discussions at public hearings on such programs had no "adversarial racial motive" and the parties involved were not persons of "bad will." "[P]ure, raw economics in competition," according to him, drove the dialogue. It made sense, for example, to locate a program in an institution with better infrastructure and that could "maximiz[e] [gains from limited] state resources."[211] Dr. Richardson interjected that an institution could not have *better* infrastructure if facilities were never built.[212] Bypassing such institutions denies them opportunities to improve their infrastructure. In turn, this diminishes the chances of placing excellent programs on their campuses. Dr. Richardson stressed that this was "a vicious cycle."[213]

Commissioner Taylor asked if any states circumvent program duplication.[214] The states of Oklahoma and Kentucky, according to Dean Pierce, successfully avoided program duplication. Kentucky State University, an HBCU, is the only state-supported institution that offers a master's level public administration program. White students seeking this degree enroll in large numbers even though Kentucky State University is an HBCU. Dean Pierce opined that if the University of Kentucky, a geographically proximate traditionally white institution, secured a master's level program in public administration, white students undoubtedly would leave Kentucky State University for the former, leading to educational apartheid. He noted that Florida A & M University, an HBCU, and nearby Florida State University, a traditionally white institution, duplicate an engineering program and wondered rhetorically which of the two institutions draws the most students.[215]

Dean Pierce stated that the State of Oklahoma located a School of Physical Therapy at Langston University, an HBCU. Most of the program's students were white. The fact that Langston University is an HBCU apparently was of little consequence since student interest lies in earning the degree in physical therapy. Dean Pierce also noted that the State of Virginia had circumvented program duplication. However, he reiterated that when funding for higher education is tight, as is the case in the States of Georgia, Maryland, Ohio, and Tennessee, the present state of affairs in Oklahoma and Kentucky might not continue to prevail. He emphasized again that the critical action occurs in the state legislatures. Depending on states' economic circumstances and available funding for higher education, state legislatures are HBCUs' "greatest threats or ... greatest champions."[216]

### *A Complementary System of Higher Education*

Against many states' backdrop of shrinking revenue sources and looming Social Security and Medicare crises—difficulties which are expected to grow over time—and the heavy dependence of public and private HBCUs on state and federal funding, Chairman Reynolds asked if HBCU officials have discussed any alternative funding models.[217]

Dr. Richardson agreed that one could not ignore the severe budget reality. In fiscally difficult times, the higher education community must deliver particularly efficient quality education. He proposed a complementary system of higher education in which institutions' high quality academic programs are not unnecessarily duplicative. This system requires states to identify "high-demand, unique programs" that are key to workforce development and strategically place them in a few geographically dispersed colleges and universities. Efforts to avoid duplication would naturally not extend to core liberal arts programs.[218]

Dr. Richardson reported that the State of Maryland initiated such a complementary higher education system for a short time in the late 1970s and early 1980s. For example, the state started an engineering program at Morgan State University in 1984 and did not immediately duplicate it in nearby white institutions. However, Dr. Richardson pointed out that even before the state had fully invested in the program to ensure growth and excellence, it created redundant programs endowed with better facilities in geographically proximate white institutions. In Dr. Richardson's judgment, an efficient solution to budgetary constraints is "greater efficiency in the way ... programs [are distributed] across [a state's] universe of institutions."[219] Commissioner Yaki commented that a current debate at the University of California (UC) system concerns programmatic specialization by campus. For example, some UC campuses would specialize in engineering, others in architecture, and additional ones in alternative disciplines. Raising his question in the context of program duplication and resource allocation, he wondered if HBCUs' administrators had similar discussions about academic specializations within their states.[220] Dr. Richardson pointed out that resource distribution within any state concerns the entire higher education system. The crux of the issue is the disparity between already underresourced black colleges and white colleges. According to him, the most efficient is a complementary or nonduplicative state system of higher education that distributes resources according to state needs.[221]

Commissioner Yaki asked about the process by which a state allocates academic programs to institutions. Specifically, he inquired whether the Commission should examine or encourage OCR to review the Title VI implications of program duplication and the allocation process.[222] Dean Pierce reaffirmed the political nature of program distribution and the critical role of state legislatures. For example, in Georgia, officials of the influential University of Georgia or Georgia Institute of Technology would lobby legislators for budget and academic programs. Representatives of HBCUs, such as Albany State University or Fort Valley State University, were seldom involved in the dialogue.[223]

Dean Pierce also argued that Title VI violations had occurred. *Adams v. Richardson* and *Adams v. Califano* cited program duplication as a remnant of segregation and as violating the federal policy of equal protection for people of color attending publicly funded institutions of higher education. He said the federal circuit court and later the Supreme Court accentuated this point in *Fordice*. Based on the holdings in *Adams* and in *Ayers*, OCR adopted a policy that similarly opposed program duplication.[224]

Dr. Richardson commented that the failure to enforce Title VI with respect to program duplication has a negative effect on desegregation in higher education. He explained that in the early 1970s, Morgan State University was "unique," being the only institution in the Baltimore area to offer several graduate programs.

Consequently, a little over 50 percent of its graduate enrollment was white. By the early 1980s, with program duplication and the state's refusal to invest in Morgan State University's graduate programs, the graduate enrollment became overwhelmingly black. [225] Dr. Richardson emphasized that state and community workforce needs should drive the establishment of new academic programs. States should locate similar programs on a small number of geographically dispersed campuses enabling students to obtain a quality education in their particular fields regardless of which of the dispersed institutions they choose to attend. In such a scenario, an institution's program -- and not the racial composition of its student population -- serves to draw students.

Such a complementary system at the state level, Dr. Richardson asserted, would dismantle a dual system of higher education.[226] He added that the State of Maryland is again investing in programs duplicative of those at Morgan State University at a close-by white college, even before it has fully funded those at the HBCU.[227]

Dr. Richardson anticipated that OCR would shortly review Maryland's progress in desegregating its public higher education system. He said that the level of desegregation at HBCUs had decreased because of excessive focus on enrolling black students in white institutions. The final measure of whether desegregation has been successful, Dr. Richardson emphasized, is if the higher education community and the state have given students "a choice to [enroll in] black ... or white institutions [that does not entail] ... sacrifice [of] the quality of [campus] life or [academic] programs."[228]

Dean Pierce noted that states remaining under federal higher education desegregation agreements are required to submit reports periodically to OCR. OCR should use those reports to monitor progress and determine states' efforts towards compliance with Title VI.[229] However, Dean Pierce was clear that OCR does not enforce Title VI with regard to unnecessary program duplication.[230]

### *Dearth of Black Male Undergraduates at TWIs*

Commissioner Kirsanow commented on the dearth of black male undergraduates at white institutions. Of note, at some institutions 80 percent of the black student population is female. The numbers of black male and female undergraduates are somewhat more balanced at HBCUs. With respect to STEM programs, however, the enrollment is disproportionately male. This is true at other institutions as well, but the gender imbalance is pronounced at HBCUs.[231]

Dr. Richardson said the absence of black males on campus, whether HBCUs or white institutions, is bewildering. At Morgan State University, 58 percent of the undergraduate students are females and 42 percent males. In addition, black males generally graduate at reduced rates compared to black females; thus, to the extent that HBCUs have increased their proportions of black males, their overall graduation rate is lower.[232]

### *Policy Initiatives to Support HBCUs*

At the culmination of the panelists' presentations and discussions with Commissioners of the educational effectiveness of HBCUs, Commissioner Kirsanow asked for suggestions of federal policy initiatives that might sustain and enhance these institutions.[233] Panelists offered recommendations in the following five areas.

### *Complementarity in Academic Programming*

Dr. Richardson recommended that the states return to complementary academic programming among institutions within a higher education system while avoiding unnecessary program duplication in geographically proximate HBCUs and their white counterparts.[234]

### *Continued Enhancement of HBCUs*

Dr. Richardson also advised states to continue to enhance HBCUs to the point of comparability with their white counterparts. Federal investment in HBCUs, for example, to build up and augment graduate programs, would make HBCUs more competitive in federal grant applications.[235] Other benefits of such investment include higher numbers of African-American students enrolling in and graduating from STEM programs.[236]

### *Enforcement of Federal Civil Rights Laws*

Dean Pierce stressed that OCR must enforce federal civil rights laws in order to ensure the continued vitality of HBCUs. Enforcement is particularly critical in seven states (Ohio, Pennsylvania, Florida, Kentucky, Maryland, Texas, and Virginia) that have been found to be in violation of Title VI of the Civil Rights Act of 1964.

Their violations have not yet been corrected. Dean Pierce believed that addressing these critical states will prompt others to examine if their own program allocations are within the law.[237]

### *Increase Title III Support for HBCUs*

Mr. Merisotis called for a substantial increase in funds for Title III of the Higher Education Act of 1965. Title III has been significant in the success of HBCUs over the last 20 years.[238]

### *Increase in Financial Aid*

Mr. Merisotis supported an increase in student financial aid, especially in the form of grants. HBCU students are "about twice as economically disadvantaged as students in other institutions."[239]

### *Threats that Imperil HBCUs*

Commissioner Kirsanow also asked panelists to identify critical threats that undermine the viability of HBCUs.[240] Panelists identified threats in the following three areas.

### *Program Duplication*

Dean Pierce emphasized that unnecessary program duplication is an arch-enemy of HBCUs.[241]

### *Declining Financial Resources*

Mr. Merisotis identified declining financial resources as a major threat to HBCUs. HBCUs are increasingly serving educationally and economically disadvantaged students. They thus require additional resources to meet the substantial academic support and financial needs of these students.[242]

***Right to Exist Questioned***

Mr. Merisotis said HBCUs are the only group of institutions whose right to exist is frequently questioned. This adds to the challenges HBCUs face daily as they strive to teach educationally unprepared and economically disadvantaged students.[243]

# STATEMENTS

## Louis W. Sullivan, M.D., Chair of The President's Board of Advisors on Historically Black Colleges and Universities

***Contributions of Historically Black Colleges and Universities to the Nation***

For more than two centuries the United States has struggled with the social and economic consequences of former systems of legally sanctioned slavery of its Black citizens in Southern states followed by legally-sanctioned segregation and discrimination, based upon race.

The majority of the nation's 106 Black colleges were created in the second half of the nineteenth century, following the emancipation proclamation by President Abraham Lincoln, which abolished slavery on January 1, 1864. Many of these colleges were created by religious organizations and others by state governments.

With passage of voting rights legislation in the 1960s and other legislation designed to eliminate the vestiges of segregation and discrimination, the question has arisen about the need for and the educational effectiveness of historically Black colleges and universities. For most young people entering college, this is their first experience away from home for an extended period. It is a time of socialization, of developing a clear identity and a time for reinforcing their values. It is during this time that young people move from the familiar protected environment of home to the new, more open, challenging and less secure ambiance of the college campus.

For some African-American young people, this interplay of academic, social and personal development which occurs on the campus of Historically Black Colleges and Universities (HBCUs) during their personal transition from home to the wider world can have a profound influence on their development as scholars, future family members, members of the country's workforce and as responsible citizens. An example of this is illustrated by the number and percentage of the graduates of some HBCUs who are successful in gaining entry to, and graduation from, schools of medicine, engineering, law and other fields. For a number of HBCUs these percentages are equal to or exceed the outcomes achieved by the African-American students and graduates from predominantly white colleges and universities which are often older and wealthier.

Among the HBCUs that have demonstrated this level of success in their graduates are Xavier University, Spelman College, Morehouse College, Florida A&M University, North Carolina A&T University, Jackson State University and others.

This phenomenon may also be seen at the professional school level. I wish to share with you our experience at Morehouse School of Medicine, (MSM), a predominantly African-American Medical School which was founded in 1975 by Morehouse College, to increase the number of African-American (and other minority) physicians in Georgia and the nation.

MSM began with modest financial resources, admitting its first class as a two-year school of medicine in 1978, becoming a four-year medical school in 1981 and receiving full accreditation of its M.D. program in 1985.

Among MSM's 800 M.D. alumni today are (1) a state health commissioner, (2) a vice president of a large medical school, (3) the personal physician of the president of South Africa and (4) physicians who are providing care in previously-underserved rural and urban areas in Georgia and around the nation.

Today, MSM's students pass national examinations given to medical students at or above the passing rates of medical students nationwide, although MSM is one of the youngest medical schools in the United States and does not yet have the level of financial resources of most U.S. medical schools.

What accounts for this experience of the students and the graduates of HBCUs? There are multiple factors, including (1) the dedication of the faculty to their teaching responsibilities, (2) the supportive social environment, (3) the strong encouragement given to the students to explore a full range of career possibilities (including leadership roles) in business, the sciences, public service, education and other fields and (4) the role models for students among the faculties of the HBCUs.

As U.S. citizens, all of us look forward to the day when the lingering vestiges of segregation and discrimination will no longer be present in our society. But, our experience as a nation has shown us that to reach that goal will require the sustained and dedicated efforts of all of us that includes the contributions of HBCUs to our nation's higher education community and the effectiveness of HBCUs in facilitating the academic, social and personal development of a significant number of our African-American (and other minority) citizens.

## Earl S. Richardson, President, Morgan State University

Since the Civil Rights Act of 1964 and the first serious efforts by the federal government to enforce that legislation beginning in the early 1970s, there has been a continuing discussion about the role that historically Black campuses should play in contemporary higher education. Before that legislation, the role of HBCUs was clear—they were open to anyone of any race or ethnicity, but they were essentially the only place that African-Americans could obtain a college education in the Southern states. Because of the high concentration of Blacks in the South, a very high percentage of the prominent African-Americans in public life and the professions nationally have obtained their degrees from HBCUs This is the case in the state in which Morgan is located; Morgan graduates have, among other accomplishments, been the first Black judge in Maryland, the first Black congressman, the first Black state senator, and the first Black official to hold any statewide office.

The Civil Rights Act changed the rules of the game. Now, Black students for the first time could attend a traditionally white campus, if they qualified. And the traditionally white schools, under legal pressure to increase minority enrollments, provided inducements, particularly financial ones, to Black students who in most instances could barely afford college, if at all. By contrast, HBCUs, by design, had been neglected. It had been the philosophy in segregated states to limit the educational opportunities available to Blacks at all levels of education, and neglect of HBCUs was part of this philosophy. Any campus president will tell you that it takes decades of hard work and adequate resources to build a quality institution. By the time the Civil Rights Act was enforced, most HBCUs were nearly a century behind their white counterparts. It is not surprising then that when the federal government began to enforce the Civil Rights Act, the HBCUs were not in a position to compete for their traditional constituency, much less for white students. When people have choices, they choose their best alternative and, in most instances, HBCUs were not attractive options.

The initial impact of federal desegregation enforcement was devastating to Black educational institutions. At the K-12 level, Black schools were so poorly funded that most closed and their staffs were terminated as the students who would historically have attended them enrolled in formerly segregated predominantly white schools. At the college level, the movement of students was almost entirely in one direction, from HBCUs to formerly segregated predominantly white campuses. Morgan lost so many Black students after the early 1970s that it has been only recently that it has again enrolled students in comparable numbers. As the white campuses developed, HBCUs also usually lost their white enrollment. At Morgan, which had non-duplicated graduate programs in the Baltimore area until the early 1970s, we went from a large graduate student body that had a slight majority of white students to one that was almost entirely Black ten years later. There is little question as to why the future of HBCUs was suspect.

However, unlike the K-12 experience, a number of factors have intervened to keep most HBCUs in business. Because initial federal desegregation efforts were focused on moving minority students to historically white campuses, HBCUs suffered enrollment declines.

Therefore, the federal government subsequently began to require states to enhance their HBCUs if they were to conclude acceptable desegregation plans. This helped many of these campuses. Second, there remained and remains today a strong core of students who want to attend an HBCU. Third, demographic trends since the since 1960s have been more favorable for the young Black population than for the young white population. This began to translate into increasing numbers of college-age Blacks in the mid-1980s, a period when the white college-age population would still be declining for another decade. While we are in a temporary period of increases for all racial groups of college age, in a few more years, the white college-age population will again decline. Finally, there has been the attraction of prestige. Whether due to rankings published in *U.S. News and World Report* and elsewhere or other factors, growing numbers of predominantly white campuses have given priority to increasing the academic profile of their entering classes. This has negatively affected Black students, who on the average, score lower on standardized tests and other measures of pre-college academic achievement than white students. HBCUs, other liberal admissions campuses, and community colleges have become their campuses of choice in recent years. For the most part, their enrollments have increased significantly as a result.

While the role of HBCUs has been of significance in providing access to the growing young minority population, measures of educational outcomes suggest a role of continuing importance. It is clear we have a long way to go and will probably have to change our priorities if we are to ever approach educational equity for Blacks and whites in our country. This goal is important not only from a social justice perspective but from the perspective of our global competitiveness. To a greater extent than at any time in our history, advanced education is important for our national well-being and we need to utilize the talents of as many of our citizens as possible in this environment.

Perhaps the single best indicator of how well we have been educating our young population is the percentage of the 25-29 age group holding at least a bachelor's degree. I am going to use 1970 as a base year because that is the period during which we began to emphasize educational equality between the races and affirmative action. This emphasis has resulted in numerous high profile programs to improve minority educational attainment. Coincidently, it is the period during which the value of a high school degree relative to a college degree began to decline. The news is not good. While higher percentages of both Black and white young adults hold college degrees than in 1970, the gap between the races has grown due to a more rapid increase in white college degree attainment. In 1970 the difference in the percentages of young Black and white adults holding a college degree was 10 percentage points. Now it is 17 percentage points, and the size of the population is much larger. This translates into a very large numerical difference. In 1970 about 160,000 more Blacks in the 25-29 age group would have had to hold a college degree to have achieved parity with the white population. Today, that gap amounts to over 400,000 persons in this age group. This means that we have to almost double the number of baccalaureates awarded to Blacks each year to achieve parity among young adults. This is a huge task.

Some might argue that we need to be patient in our expectations for equality in degree attainment. We have to start at the lower grade levels. The news is not good here, either. On most measures of educational progress at the K-12 level, the gap has not been closing. Over the past decade white SAT scores have risen more rapidly than Black SAT scores, even though they were much higher to begin with. While I would agree that there is a great deal of potential for improving the qualifications of minority students in the educational pipeline, I also would argue that we cannot wait. We have been waiting for over three decades without visible signs of progress. Instead we need to be concerned about how we can take college-age students where they are and improve their chances of obtaining a degree.

To illustrate how pre-college academic indicators affect admissions at selective colleges it is useful to look at how many students score in various ranges on the SAT. Most people know that whites score some 200 points higher on the combined math and verbal portions of the SAT. At the upper ranges, where majority campuses increasingly recruit, the differences are stark. Whereas 6 percent of whites score 1400 or higher on the SAT M+V, fewer than 1 percent of Blacks do so. Fewer than 5 percent of Blacks score 1200 or higher compared to 24 percent of whites. This is the range in which both moderately and highly selective campuses recruit. Fewer than one-quarter of Blacks score over 1000 on the SAT while 60 percent of whites do so. It is no wonder that the lack of lower income and minority students at selective campuses has been the topic of so many articles and conferences in recent years. These campuses simply cannot mirror the increasingly diverse population of the nation if they remain so selective. Under current circumstances, they would have to tolerate a large disparity in pre-college credentials between the races to achieve anything approaching a student body that is representative of the college-age population. *This is why the selective campuses recruit from each other's pool rather than expanding the size of the pool of minority students enrolling in college*. Hence, less selective campuses, including HBCUs, offer the only realistic option for significant increases in the numbers of minority students enrolling in and graduating from college.

A useful rule of thumb is that campuses that are relatively selective graduate student bodies that have only about half the percentage of African-Americans as the relevant population from which they recruit. For example, the Ivy League campuses, as a group, are national universities and have a graduating class that is 7 percent Black. This is about half the national representation of Blacks. UCLA and Berkley have graduating classes that are only 4 percent Black, half the African-American representation in the state's population. The University of North Carolina – Chapel Hill graduates a class that is 11 percent Black, about half the Black representation in that state. Hence, there is little that such campuses can do to significantly increase the representation of Blacks among degree holders nationally.

One might also argue that there are many minority college students who could be admitted to more selective institutions if non-traditional measures of academic potential were utilized. The facts show a different picture. In most states, once you look beyond the flagship campus(es), you see relatively low graduation rates for African-Americans. Morgan, for example, has a graduation rate of 43 percent. While this sounds like it is just average, it is well above average for African-Americans at urban universities across the country. Hence, it is unlikely that selective campuses are likely to double the percentage of African-Americans in their graduating classes by recruiting from the pool from which less selective campuses in their states recruit.

This leads one to look at the productivity of liberal admissions campuses as a source of African-American graduates. The HBCUs stack up well on this measure. For example, 24 HBCUs graduate as many Black students as the University of Michigan and the University of North Carolina-Chapel Hill. My campus graduates almost as many Black students as all of the Ivy League campuses combined. Southern University graduates more. Howard University and Florida A & M graduate more Black students that the total University of California System, the largest public system in the nation. Thus, because the size of the degree attainment gap is so large and HBCUs are quite productive in the overall picture, they should be an integral part of any strategy for increasing the number of graduates nationwide.

One reason that HBCUs are effective in increasing the number of minority graduates is their tradition of providing a nurturing environment. These campuses always have had to educate many students who are not as well prepared for college as is desirable. For different reasons than in the past, there remains a large pool of under-prepared minority students of college age. A support system that is part of the campus culture, rather than a programmatic appendage, is a real advantage in addressing student academic needs, particularly when relatively large numbers may be in need of extra assistance. About 10 years ago, Morgan was one of two HBCUs in a national study in which gains in student achievement were measured at various points during college. Even though a number of the campuses in the study were considerably more selective than the two HBCUs, the gains in educational achievement were greater for African-Americans at the HBCUs than at the other campuses. The researchers conducting the study attributed this to the supportive environment found at the HBCUs.

On other measures of educational outcomes our campus does well even though we are an access-oriented institution whose graduates must compete with other graduates who typically began their college careers with stronger records of preparation. Our annual follow-up surveys of graduates find that the employment measures of our graduates are equal to statewide averages. Our graduate and professional school going rates are consistently above the state average. We are one of the leading producers of Fulbright scholars in the region, and near the top nationally on a size-adjusted basis.

While I think that Morgan as well as other HBCUs do a pretty good job of making students from difficult backgrounds into productive middle class citizens I think that a lot of other access-oriented campuses do so as well. I include most regional college and universities, urban institutions, and community college in this group. However, the general public and many in the higher education community associate effectiveness with having a wellcredentialed entering freshman class. I would submit that the intellectual distance your students travel while they are enrolled at your campus is a better measure. I think that until we face the fact that our current strategies have not closed the educational achievement gap between the races, many will continue to be preoccupied with input rather than output.

In summary, the magnitude of the educational attainment gap is large. The trend in recent decades has been a widening of this gap. Campuses which are relatively selective do not increase the size of the pool of African-Americans going to college. Rather they recruit from the relatively small pool of students who would be successful wherever they enrolled. Access-oriented campuses such as HBCUs have the greatest impact on degree attainment for students of all races. It is these campuses where additional resources will have the most impact because they have the ability to increase the size of the pool of students enrolling and completing college.

## Jamie P. Merisotis, President, Institute for Higher Education Policy

Mr. Chairman and Members of the Commission:

Thank you for this opportunity to appear at this Commission briefing regarding the educational effectiveness of Historically Black Colleges and Universities (HBCUs).

I would like to begin with a brief introduction to the Institute for Higher Education Policy and our role in the policy process. Established in 1993, the Institute is a non-profit, non-partisan research and policy organization whose mission is to foster access and success in postsecondary education through public policy research and other activities that inform and influence the policymaking process. The Institute's work addresses an array of issues in higher education, ranging from technology-based learning to quality assurance to student success. However, the Institute is probably best known for its studies and reports concerning higher education financing at all levels. These studies and reports address topics ranging from federal and state student financial aid to state funding formulas to trends in institutional expenditures and revenues. We also have worked on higher education financing issues in the international context, especially in Southern Africa and Eurasia.

The Institute's independent voice on these issues is well known. Our primary funding is derived from major foundations that are interested in supporting higher education research and analysis. We also have conducted a fair amount of analytic work at the behest of state governing and coordinating boards for higher education, as well as national governments outside the U.S.

In addition to this independent analytic work, the Institute also plays a unique role as facilitator and convener of the Alliance for Equity in Higher Education, a groundbreaking collaborative established in 1999 by the American Indian Higher Education Consortium (AIHEC), the Hispanic Association of Colleges and Universities (HACU), and the National Association for Equal Opportunity in Higher Education (NAFEO). Combined, these three organizations represent over 350 colleges and universities that serve more than one-third of all students of color in the United States. As institutions educating the nation's emerging majority populations, these minority-serving institutions (MSIs) of higher education represent the vanguard of the nation's future potential and promise.

As the Alliance's facilitator, the Institute has served in two key roles: we have conducted research that examines the unique context of MSIs in areas such as teacher education and the use of technology, and we have functioned as a manager of programs that address common issues of concern to MSIs. Our program management responsibilities include a program to train future presidents of HBCUs and other MSIs (MSI Leadership Fellows), supported by the W.K. Kellogg Foundation; a program to enhance the data and analytic capacities of MSIs committed to increasing student learning and success (Building Engagement and Attainment of Minority Students [BEAMS]), supported by Lumina Foundation for Education; and an initiative to disseminate knowledge learned as the result of a decade-long investment to improve undergraduate education in Science, Technology, Engineering, and Mathematics (Model Institutions for Excellence), jointly supported by the National Science Foundation and the National Aeronautics and Space Administration. The Alliance also recently has assumed leadership and management of the National Articulation and Transfer Network (NATN), a major national initiative designed to increase enrollment, retention, and graduation rates for minority students.

Improving the educational effectiveness of higher education continues to be one of the most important contributions that the federal government, states, individuals, and the private sector can make to our national well-being. The simple fact remains that increasing educational opportunities for all Americans results in tremendous public private social and economic benefits. As the chart at the conclusion of my testimony points out, going to college is much more than just a process of enhancing one's personal economic status. The combination of societal and individual benefits of higher education must continue to motivate what we do at many levels, and I hope it will be an important consideration for the Commission as it takes up this important issue of the educational effectiveness of HBCUs.

At this briefing you will be hearing from many distinguished leaders from the community of HBCUs, individuals who speak with a great deal of authority and experience. I will not take time to cover territory about the historical and contemporary role of HBCUs as key drivers of educational effectiveness, since these other witnesses are highly qualified to do so. Instead, as a complement to their testimony, I will focus my remarks on several issues regarding the effectiveness of HBCUs that draw from the Institute for Higher Education Policy's combined work as both an independent research and analytic organization and as an organization that manages programs on behalf of HBCUs and other minority-serving institutions.

My remarks address four areas concerning the educational effectiveness of HBCUs. These are somewhat lesser known, but nonetheless critically important, ways in which we should be viewing the educational effectiveness of HBCUs. They are:

- HBCUs as leaders in student engagement;
- HBCUs as community-based institutions that promote civic engagement and service learning;
- HBCUs as drivers of educational attainment for low-income, first generation, and disabled students; and
- HBCUs as examples of success in the national effort to improve the quality of science, technology, engineering, and mathematics (STEM) education and research.

In each case, I will point to a specific example of success of the effectiveness of HBCUs in this area. This information is illustrative of the types of accomplishments we have seen at HBCUs in recent years.

***HBCUs and Student Engagement***

One of the most important trends in higher education in the last decade has been an effort to document how well institutions engage in effective educational practices—that is, activities that are empirically related to desired learning and personal development outcomes of college. Perhaps the best example of this is the work that has been conducted by the National Survey of Student Engagement (NSSE), administered by the Indiana University Center for Postsecondary Research. NSSE is designed to obtain information from colleges and universities nationwide about student participation in programs and activities that institutions provide for their learning and personal development. The results provide an estimate of how undergraduates spend their time and what they gain from attending college. Nearly 1,000 higher education institutions have administered NSSE since it began national administration in the year 2000.

The NSSE survey questions focus on the frequency with which students participate in effective educational practices. For example, students are asked how often they discussed ideas with faculty outside of class, participated in a community-based project as a part of a course, and used an electronic medium for an assignment. In addition, the survey asks students to identify the degree to which their courses emphasize different mental processes (such as memorizing and analyzing), how many hours per week they spend studying, working, or participating in co-curricular activities, as well as how they would characterize their relationships with people on campus. All of these factors have been shown through prior research to influence student retention and graduation.

According to 2004 and 2005 data collected by NSSE at 37 HBCUs, African-American students at HBCUs report more frequent interactions with faculty than African-American students at predominantly white institutions (PWIs). HBCUs also generally appear to provide more supportive learning environments for students, including more contact with faculty. Students at these HBCUs report a greater belief that their institutions contribute to their personal spiritual growth, and report a higher likelihood that they will vote compared to their counterparts at PWIs.

These data from NSSE, combined with a significant body of research undertaken by others, suggest that HBCUs provide a superior level of student engagement, and therefore offer an educational experience that enhances the intellectual gains and accomplishments of students. This increased engagement of students is an important indicator of the effectiveness of HBCUs that deserves closer examination and study by those seeking to enhance the payoffs of the investment in HBCUs.

***HBCUs and Civic Engagement/Service-Learning***

A key national issue is the extent to which institutions of higher education contribute to civic engagement and participation in our democratic institutions, such as voting, volunteering, and community involvement. In general, we know that higher educational attainment is highly correlated with increased civic engagement. For example, a 50-state study of the benefits of higher education published last year by the Institute for Higher Education Policy found that 36 percent of Americans over the age of 25 with a bachelors degree volunteer, compared to just 21 percent of those with a high school diploma. Similarly, voting rates in national elections for individuals with a bachelor's degree are nearly 50 percent higher than for those with a high school diploma. Clearly higher education makes a profound difference in terms of our national civic well-being.

The question for higher education institutions is what specifically they may be doing to foster those values and goals. While there are many ways in which colleges and universities can contribute to the development of civically-engaged citizens, one way is through the use of service-learning or other activities that integrate service and democratic engagement into the curriculum of the institution.

According to the National Campus Compact, which represents over 950 colleges and universities committed to the civic purposes of higher education, HBCUs and other MSIs do a remarkable job of civically engaging students. The 2004 Campus Compact membership survey found that these institutions are more likely than others to require service and service learning for graduation. They also found that HBCUs and other MSIs are more likely than other colleges and universities to have a community service or service-learning office; to have a director of community service or service learning on their campus; and to have partnerships with local K-12 schools and faith-based organizations.

These intentional strategies by HBCUs to engage their students at the community or civic level are not well documented in the national literature about service learning and civic engagement, but clearly deserve further exploration. As an example of the effectiveness of institutions, they point to a profoundly different approach to student success than what is seen at many other colleges and universities.

### *HBCUs and Success for Low-Income, First Generation, and Disabled Students*

Another indicator of the effectiveness of HBCUs is the degree to which they have committed to the educational advancement of low-income, first generation, and disabled students. HBCUs are well known for the opportunities they provide to students who come from educational and economically disadvantaged circumstances. In so doing, these institutions work hard to provide these students with additional support, guidance, and mentoring that will improve their opportunities to get into, and succeed in, college. For example, the federal government has long supported increased opportunity for these populations through the federally funded TRIO programs. These programs, authorized under the Higher Education Act, provide a continuum of services from pre-college to pre-graduate level study for the nation's low-income, first-generation, and disabled students. In FY 2006, the $828 million in funding for TRIO programs supported more than 850,000 students in over 2,700 distinct TRIO programs.

There are a total of seven TRIO programs. The pre-college programs include Talent Search, which provides counseling and information about college admissions requirements and student financial aid to young people in grades six through 12, and Upward Bound, which works with students starting in the 9th grade to provide instruction in literature, composition, mathematics, and science on college campuses after school, on Saturdays and during the summer. In addition, Upward Bound Math Science helps students from low-income families to strengthen math and science skills, frequently providing research opportunities for underrepresented students, while Veterans Upward Bound provides intensive basic skills development and short-term remedial courses for military veterans to help them successfully transition to postsecondary education.

At the college level, Student Support Services programs provide tutoring, counseling, and supplemental instruction to help students stay in college through the completion of a degree, and in the case of community colleges, assist them in the transfer process. The Ronald E. McNair Post-Baccalaureate Achievement programs are designed to encourage low-income students and minority undergraduates to consider careers in college teaching as well as prepare for doctoral study. Students who participate in this program are provided with research opportunities and faculty mentors. Finally, the Educational Opportunity Centers located throughout the country primarily serve displaced or underemployed workers from low-income families.

According to the Council for Opportunity in Education, nearly three quarters of all HBCUs have TRIO programs, serving nearly 70,000 students. This compared to less than one quarter of all other colleges and universities. The more than $70 million in support provided by these programs to serve students at HBCUs goes a long way toward increasing the odds of student success than students who do not have the benefit of these programs.

### *HBCUs and Science, Technology, Engineering and Math (STEM)*

The final example of ways in which HBCUs are educationally effective is the role that they play in improving the quality of STEM education and research. At the national level, investments in science, technology, engineering, and mathematics have been universally accepted as a national imperative. The President's proposed 2007 Budget, for example, advocates significant new investments in these efforts as key drivers of the nation's global competitiveness and economic capacity. However, research indicates that African-Americans are significantly underrepresented both as a percentage of the national STEM workforce and as a proportion of those enrolling and succeeding in STEM programs at colleges and universities.

An interesting example of a program that is designed to increase the quality of STEM education and research at HBCUs is the Historically Black Colleges and Universities-Undergraduate Program (HBCU-UP) at the National Science Foundation (NSF). HBCU-UP provides an array of support to institutions to help strengthen STEM education and research. This includes curriculum enhancement, faculty professional development, undergraduate research, academic enrichment, and infusion of technology to enhance STEM instruction, collaborations with research institutions and industry, and other activities that meet institutional needs.

According to NSF, Math gatekeeper passing rates—including in Algebra, Pre-Calculus, and Calculus I—have improved at 14 HBCU-UP grantee sites that have had projects in place for five years. Improvements have also been seen in other STEM gatekeeper courses, such as Biology I and Physics I. These improvements correlate to the faculty development and course revision and enhancement activities that are implemented by HBCU-UP projects. Similarly, several indicators of research and professional development activity have increased as a result of HBCU-UP—such as the number of refereed journal articles and other publications, conference presentations, and proposals submitted for external funding. This reflects increased research opportunities at HBCU campuses for undergraduates to experience either in their classes or as researchers. Approximately 25 percent of STEM graduates from these HBCUs now have had an undergraduate research experience that better prepares them for success in graduate school.

These brief examples of the educational effectiveness of HBCUs are not intended to be definitive or inclusive. Rather they are designed to illustrate that the educational effectiveness of HBCUs has many dimensions that go beyond the simple calculation of aggregate graduation rates, retention rates, or job placements. In assessing the educational effectiveness of HBCUs, it is critical to consider a wide array of information and data that paint a more complete portrait of effectiveness than might be indicated by more narrowly drawn measures. This more comprehensive picture of educational effectiveness can then be used to provide a fair assessment of HBCU performance and, in so doing, help to improve the targeting of strategies to continuously upgrade quality and performance at these nationally essential institutions of higher learning.

Thank you again for this opportunity to appear before the Commission on this important issue. I would be pleased to answer any questions you may have.

**The Array of Higher Education Benefits**

| Econo-mic | Public | Private |
|---|---|---|
| | *Increased Tax Revenues | *Higher Salaries and Benefits |
| | *Greater Productivity | *Employment |
| | *Increased Consumption | *Higher Savings Levels |
| | *Increased Workforce Flexibility | *Improved Working Conditions |
| | *Decreased Reliance on Government Financial Support | *Personal/Professional Mobility |
| | *Reduced Crime Rates | *Improved Health/Life Expectancy |
| | *Increased Charitable Giving/Community Service | *Improved Quality of Life for Offspring |
| Social | *Increased Quality of Civic Life | *Better Consumer Decision making |
| | *Social Cohesion/Appreciation of Diversity | *Increased Personal Status |
| | *Improved Ability to Adapt to and Use Technology | *More Hobbies, Leisure Activities |

## Raymond C. Pierce, Dean and Professor of Law, North Carolina Central University

### *Failure of the Federal Government to Enforce Federal Civil Rights Laws: A Real Threat to Historically Black Colleges and Universities*

The value of Historically Black Colleges and Universities (HBCU) to our nation is clear. Our nation or any nation benefits from an educated population. HBCUs continue to provide educational opportunities for African-Americans in significant numbers. There is no indication that closing public HBCUs would create a comparable shift in African-American student enrollment in and graduation from traditionally white institutions. The case has been made for HBCUs in terms of their effectiveness and contribution toward the education of the people of this nation. This presentation, however, focuses on a real and continuing threat to HBCUs in large part due to the federal government's refusal to enforce federal civil rights laws as they relate to African-Americans attending HBCUs. This threat puts in jeopardy the significant contribution HBCUs provide in allowing our nation to be competitive in a world where higher education is necessary for participation in a fast moving global economy.

For the most part federal civil rights activity in higher education involving HBCUs followed school desegregation litigation in the wake of *Brown v. Board ofEducation*. In the area of higher education, the *Adams* cases of the 1970s tied equal protection for African-Americans in the Southern and border states to state treatment of publicly funded HBCUs. Nineteen states were involved in what can be called higher education desegregation activity. Higher education desegregation actually involved legal actions brought by private individuals seeking equal funding for HBCUs in relation to traditionally white institutions. These actions were coupled with federal efforts through intervention to desegregate state systems of higher education that were created with an apartheid approach to providing public higher education.

In the *Adams* cases Southern and border states with public HBCUs were determined by a federal agency (the Office for Civil Rights of the Department for Health Education and Welfare) to be in violation of Title VI of the 1964 Civil Rights Act. Essentially these states continued to operate HBCUs with resources and standards substantially below traditionally white institutions. Private individuals (Adams) alarmed by the executive branch of the federal government's refusal to address violations of federal civil rights laws turned to the federal courts. The *Adams* litigation involved actions against the Office for Civil Rights (OCR) for its failure to enforce federal civil rights laws. States found in violation of federal civil rights laws continued to receive federal funding to support state systems of higher education that were wrought with vestiges of the system of apartheid. The federal government's OCR was not requiring these states to submit plans to bring themselves into compliance with Title VI, nor was OCR crafting guidelines for any such plans. The result of the *Adams* cases was a finding against OCR, the development of federal guidelines for desegregating state systems of higher education, and Southern and border states entering into desegregation plans (Agreements) with the federal government designed to bring about compliance with Title VI.

The Agreements with the various states designed to bring about Title VI compliance was based on 1978 federal civil rights policy derived in part from the court decisions in *Adams*. The federal civil rights policy for desegregating state systems of higher education was basically in two parts: 1. Strengthening HBCUs and 2. Affirmative Action at Traditionally White Institutions (TWIs).

Strengthening or enhancing HBCUs was aimed at accomplishing two things: providing expanded educational opportunities for African-American students and attracting white students to form a more diverse student population. The history of HBCUs includes significant constriction in the educational programming offered at these institutions. Further, the physical facilities were limited consistent with the limited educational programs. The federal policy for desegregation included the placement of attractive educational programs at HBCUs to address the history of limited educational programming and thus offer expanded educational opportunities to African-Americans attending these institutions. In addition, this enhancement was intended to attract a more diverse student population seeking quality educational programs. The design was for students to select public colleges based on educational offerings and not solely on a legacy of racial congregation.

The second component of the federal higher education desegregation policy was one of affirmative action at the Traditionally White Institutions. The affirmative action here was not affirmative action in admissions but affirmative action in recruitment and affirmative action in scholarships, both targeting African-American students. The intent was to attract more African- American students to become enrolled in the public Traditionally White Institutions.

Enhancing HBCUs to provide expanded educational opportunities and attract a more diverse student population combined with affirmative action programs at TWIs designed to attract African-American students together formed the policy intended to do away with white colleges and Black colleges and create one system of colleges. That was the plan.

Most of the 19 states entered into compliance plans. Five states did not enter into agreements. Four of these states proceeded to federal court litigation. One state, Ohio, was prepared for litigation but pleadings prepared by the U.S. Department of Justice were never filed in court. Louisiana, Tennessee, Alabama and Mississippi all went to court with the Mississippi case going all the way to the U.S. Supreme Court. It was the Supreme Court decision in the Mississippi case that would produce the foundation for new federal civil rights policy on higher education desegregation that was published in 1994.

In 1988, the U.S. Secretary of Education, William Bennett, directed OCR to conduct reviews of those states that had entered into the Agreements and make determinations as to whether or not those states had brought themselves into compliance with Title VI. Using a checklist analysis OCR concluded that eight of the 19 states had performed according to the Agreements and that those states were now in compliance with Title VI. Reviews were not conducted for the remaining six states at that time (Virginia, Pennsylvania, Florida, Maryland, Texas and Kentucky). In 1994 OCR published new federal policy for higher education desegregation that was based on the U.S. Supreme Court decision in *Ayers* (the Mississippi case). The 1994 policy created a higher standard for compliance. Whereas the 1978 policy was based on a checklist analysis (e.g., did the states enhance the public HBCUs and implement affirmative action procedures at the TWIs), the 1994 policy was based on a vestiges analysis which places on states an affirmative duty to remove all vestiges of the past practice of segregation that have present day effects to the greatest extent practicable.

Pursuant to the higher standard contained in the 1994 policy the remaining six states plus Ohio entered into new five year plans (New Agreements) designed to bring about Title VI compliance. In addition, there was some debate within OCR as to whether or not the eight states that were closed out in 1988 under the old 1978 standard should be reviewed for Title VI compliance using the newer standard contained in the 1994 policy. The five year plans focused mainly on further enhancement of the public HBCUs in the seven states.

The problem we are faced with today is almost a re-visitation of the situation that led to the *Adams* cases in the 1970s. There is substantial indication that many of the seven states have ceased performance pursuant to the New Agreements. In some of the states (Ohio in particular) there is actually action being taken to revert back to a policy of constraining HBCUs in their ability to offer attractive educational programs. Clearly in many cases in these states there is no focus on compliance with federal civil rights laws. In addition, there is strong evidence that HBCUs in some of the states that were closed out in 1988 are being negatively impacted by state actions that are in direct contradiction of federal higher education desegregation policy. The most egregious of these state actions is unnecessary program duplication. Duplication of programs in colleges within close proximity was historically done for apartheid purposes. Courts and federal civil rights policy have cited the unnecessary duplication of programs as being segregative and violative of Title VI. Unnecessary program duplication in a highly competitive business of higher education poses serious threats to HBCUs particularly with respect to their ability to compete in the arena of masters and doctorate level programs.

There is every indication that OCR will continue its non-enforcement of the Agreements of which it entered into with the various states for Title VI compliance purposes. At the same time it is clear that states will continue with initiatives, primarily at the legislative level, that encroach upon the vitality and strength of HBCUs. This is all going on while at the same time federal funds are being provided to these very states that continue to have outstanding violations of federal civil rights laws. This is in many respects the situation that existed that ultimately launched the *Adams* cases.

I submit that litigation against the federal government's OCR be initiated similar to that litigation that formed the basis of *Adams*. As in *Adams*, declaratory and injunctive relief should be sought to both obtain a statement from the federal courts as to the constitutional obligations of the states in this matter and to also prevent further state actions that will have damaging effects on HBCUs. Federal court involvement is also needed as in *Adams* to declare the responsibility of OCR in this matter and to direct that agency to conduct compliance reviews of the various states with respect to the Agreements and in light of federal civil rights and higher education desegregation policy. There is some debate as to legal strategies to pursue such claims, including whether or not case law subsequent to Adams would even allow such causes of action. These issues must be resolved and the appropriate litigation commenced. Otherwise there is no real preventative measure against what clearly are real threats to higher educational opportunities for significant numbers of African-Americans.

## Mikyong Minsun Kim, Associate Professor of Higher Education, Graduate School of Education and Human Development, The George Washington University

Thank you for the opportunity to testify before the Commission. My name is Mikyong Minsun Kim, and I am an Associate Professor of Higher Education at The George Washington University. Former posts include faculty positions at the University of Arizona–Tucson and the University of Missouri – Columbia. I have also served as a grant panelist and consultant for the National Science Foundation.

While I have a wide range of scholarly interests, a great deal of my focus has been on the impact and effectiveness of Historically Black Colleges and Universities (HBCUs) on African-American students. Therefore, my testimony will focus on research findings concerning the effectiveness of these institutions. (As a possible point of interest, I have studied the effectiveness of women-only colleges as well.)

First, I will briefly compare the institutional and student characteristics of HBCUs with those of Historically White Colleges and Universities (HWCUs). Second, I will review the findings of my studies on the effectiveness of HBCUs and compare my findings with those of other studies. Third, I will discuss how HBCUs contribute to the learning opportunities of African-American students and the higher education community. Fourth, I will discuss whether and why HBCUs merit strong support.

### *Institutional and Student Characteristics at HBCUs versus HWCUs*

There are 103 HBCUs in this country (National Center for Education Statistics [NCES], 1996, 2003), and about 30 percent of the BA degrees earned by African-Americans annually are awarded by the 89 four-year (41 public and 48 private) HBCUs (NCES, 1996, 2003). Among African-American college graduates, a high percentage of political leaders, lawyers, doctors, and PhD recipients have graduated from HBCUs (Jackson, 2002; Willie and Edmonds, 1978).

Nevertheless, HBCUs have relatively fewer resources than HWCUs. HBCUs, on average, tend to have academically less prepared students and poorer institutional resources than HWCUs (e.g., physical facilities, expenditure per FTE student, and average faculty salary), and students' parents are less affluent. The findings of students' poor academic preparation and low parental socioeconomic status are consistent with Allen's findings (1992).

On the other hand, HBCUs tend to have a lower student-faculty ratio, lower enrollment, and somewhat higher student-faculty interaction, all of which are positive predictors of student development in general (Pascarella and Terenzini, 1991, 2005; Astin, 1993). From the mean comparison of Table 1, African-American students are more likely to be involved in faculty's research projects at HBCUs, and this involvement is a positive predictor for degree completion. The information in Table 1 was drawn from my recent HBCU study, published in *Research in Higher Education*.

### *Effectiveness of HBCUs versus HWCUs*

Let me share with you some of my findings. I used national longitudinal data sets (from the Higher Education Research Institute at UCLA), institutional effectiveness models, and rigorous multi-level statistical techniques (hierarchical linear and non-linear modeling) for the design and analysis of my HBCU studies. Hierarchical linear and non-linear models have well-established methodological advantages over standard regression techniques for evaluating the effectiveness of schools and colleges and handling multi-level nested data sets (Raudenbush and Bryk, 2002).

Initially, the finding of no significant difference between HBCUs and HWCUs throughout three academic outcomes (overall academic ability, writing ability, and math ability) was rather surprising. I found the same pattern of "no significant difference" between HBCUs and HWCUs in their graduates' early career earnings as well as the probability of obtaining a baccalaureate degree. In what follows, I will explain how my findings are consistent with or contrary to the previous studies on HBCUs.

### *Academic Development and Success*

My research finding that attending HBCUs is not more beneficial in developing Black students' overall academic ability, writing ability, and math ability than attending White institutions is somewhat consistent with previous studies by Central, Linn, and Parry (1970), Bohr, Pascarella, Nora, and Terenzini (1995), and Pascarella, Edison, Nora, Hagedorn, and Terenzini (1996).

The finding of no differential effect of HBCUs on obtaining a bachelor's degree is somewhat inconsistent with previous studies (Cross and Astin, 1981; Pascarella, Smart, Ethington, and Nettles, 1987; and Ehrenberg and Rothstein, 1993). Although the degree completion rates for African-American students are 55 percent for HBCUs and 63 percent for

HWCUs, college GPAs of African-American students did not differ between the two institution types.

***Early Career Earnings***

My findings show that HBCUs are doing as well as HWCUs in producing African-American graduates who are financially successful—at least in the early part of their careers. The finding of no difference in HBCUs impact on their graduates' early income is consistent with that of Pascarella, Smart and Stoecker (1989) and with that of Ehrenberg and Rothstein (1993), but it is contrary to Constantine's study (1995), which found significant economic benefits associated with attendance at HBCUs. It is also contrary to Baratz and Ficklen (1983), who found that graduates of HBCUs have lower earnings than their counterparts at HWCUs. The finding is also partly contrary to Solnick (1990), who reported that HBCU graduates started with high salaries but received smaller wage increases and fewer promotions than HWCU graduates.

We should take the finding of "no significant difference" as a positive sign that African-American students, as a group, now benefit equally in their academic success and early career earnings whether they attend HBCUs or HWCUs. By way of explanation, I speculate that the discriminatory climate at HWCUs might have eased since the desegregation movement.

***How Do Students' Learning Opportunities Differ between HBCUs and HWCUs?***

African-American students at HBCUs are more actively and deeply involved in the academic community than are their counterparts at HWCUs. As the previous studies contend, HBCUs seem to provide a more academically supportive and engaging environment for African-American students. One of my studies also indicates less satisfying and more difficult academic experiences among African-American female students at HWCUs. Although HWCUs provide African-American students with equal access (or even preferential treatment) in admissions, they may still be less likely to include African-American students in their academic communities. For example, my previous study of HBCUs revealed that helping more undergraduates participate in professors' research projects is a good strategy for improving African-American students' retention and graduation rates. The same study also suggested that African-American students had many more chances (1.5 times) to participate in professors' research projects at HBCUs. Judging by the findings of my studies, as well as those of previous investigators' studies, there are obvious compensating factors, and the two types of institution contribute to student learning in different ways: HWCUs provide more visible monetary resources and prestige, while HBCUs offer greater humane support and deeper involvement.

***Conclusion and Recommendation***

How are HBCUs able to have such an impact with relatively few resources and with students whose pre-college performance was poorer than that of their counterparts at HWCUs? How they manage to produce the same level of student outcomes as HWCUs in spite of poorer academic and financial resources needs to be investigated further in future studies. Based on numerous research findings, HBCUs appear to be more cost-effective in achieving their mission of educating Black students. Given that HBCUs are significantly underfunded relative to HWCUs, the findings of my studies and of other reports lend support to the proposition that HBCUs contribute significantly to higher education in this country and merit strong support.

Thank you for this valued opportunity to participate.

**Table 1. Comparing Student and Institutional Characteristics of HBCUs and HWCUs: Means, Standard Deviations, and Correlation Coefficients**

| | HWCU | | HBCU | | r with Black college |
|---|---|---|---|---|---|
| | Means | SD | Means | SD | |
| Student characteristics | | | | | |
| Degree completion | 0.63 | 0.48 | 0.55 | 0.50 | –0.08 |
| Respondent's early earnings | 4.23 | 1.76 | 3.97 | 1.65 | –0.08 |
| Parental income | 6.96 | 3.28 | 6.25 | 3.21 | –0.11* |
| High school GPA | 5.55 | 1.67 | 4.56 | 1.69 | –0.29 * |
| College GPA [2] | 3.67 | 0.95 | 3.79 | 1.04 | 0.06 |
| Institutional characteristics | | | | | |
| Global characteristics | | | | | |
| Selectivity in admission | 1105.00 | 120.63 | 724.80 | 77.32 | –0.73 * |
| Total enrollment | 14486.94 | 10864.51 | 1823.40 | 1002.92 | –0.67 * |
| Internal characteristics | | | | | |
| Student-faculty ratio | 20.88 | 5.76 | 15.80 | 2.78 | –0.42 * |
| Percentage: research project with faculty | 22.96 | 7.23 | 31.89 | 5.77 | 0.50 * |
| Student-faculty interaction | 7.92 | 0.18 | 8.06 | 0.16 | 0.27 * |
| Average faculty salary | 42751.94 | 6579.04 | 26814.00 | 5470.87 | –0.68 * |
| Percentage: faculty with a Ph.D. | 84.53 | 6.19 | 62.80 | 12.20 | –0.68* |
| Percentage: total instruction-related expenditure | 73.13 | 11.61 | 72.35 | 8.80 | –0.03 |
| Instruction-related expenditure per FTE student | 9246 | 5306 | 6506 | 1208 | –0.25* |
| Percentage: students transferring | 12.25 | 7.48 | 15.71 | 6.41 | 0.22 * |
| Percentage: undergraduates receiving aid | 53.53 | 13.18 | 84.00 | 11.10 | 0.67* |

Note: [1] The correlation coefficients are Spearman's rho; * $p < 0.05$ (two-tailed).

[2] The information of the variable was much reduced, and the available samples of college GPA for mean, SD, and correlation analyses were only 280.

Most of the comparative data were drawn from Kim (2002) and Kim and Conrad (2006).

Power Point Presentation

# The Effectiveness of Historically Black Colleges and Universities

United States Commission on Civil Rights
Washington D.C.
May 5, 2006

Testimony of Mikyong Minsun Kim
Associate Professor of Higher Education
Graduate School of Education and Human Development
The George Washington University

1

## Demographic Information

- There are 103 Historically Black Colleges and Universities (HBCUs) in the U.S.
- About 30 percent of the BA degrees awarded to African American students are from the 89 four-year (41 public and 48 private) HBCUs (NCES, 1996, 2003).
- Among African American college graduates, a high percentage of political leaders, lawyers, doctors, and PhD recipients have graduated from HBCUs.

2

## Institutional and Student Characteristics

- HBCUs tend to have academically less prepared students and poorer institutional resources than HWCUs.

- HBCUs tend to have a lower student-faculty ratio, lower enrollment, and somewhat higher student-faculty interaction.

3

## Institutional and Student Characteristics

- African American students are more likely to be involved in faculty's research projects at HBCUs (almost 1.5 times).

- The degree completion rates for African American students are 55% for HBCUs and 63% for HWCUs, but college GPAs of African American students did not differ between the two institution types.

4

## Effectiveness of HBCUs

- No significant difference between HBCUs and HWCUs throughout three academic outcomes (overall academic ability, writing ability, and math ability).

- The same pattern of "no significant difference" between HBCUs and HWCUs in their graduates' early career earnings as well as the probability of obtaining a baccalaureate degree

5

## Effectiveness of HBCUs

- ***Academic development.***

  My finding that attending HBCUs is not more beneficial in developing Black students' overall academic ability, writing ability, and math ability than attending White institutions is somewhat consistent with previous studies by Centra, Linn, and Parry (1970), Bohr, Pascarella, Nora, and Terenzini (1995), and Pascarella, Edison, Nora, Hagedorn, and Terenzini (1996).

6

## Effectiveness of HBCUs

- ***Degree completion.***

  The finding of no differential effect of HBCUs on obtaining a bachelor's degree is somewhat inconsistent with previous studies (Cross and Astin, 1981; Pascarella, Smart, Ethington, and Nettles, 1987; Ehrenberg and Rothstein, 1993).

7

## Effectiveness of HBCUs

- ***Early career earnings.***

  The finding of no difference in HBCUs impact on their graduates' early income is consistent with that of Pascarella, Smart and Stoecker (1989) and with that of Ehrenberg and Rothstein (1993), but is contrary to Constantine's study (1995) and Baratz and Ficklen (1983).

8

## Effectiveness of HBCUs

- We should take the finding of "no significant difference" as a positive sign that African American students, as a group, now benefit equally in their academic development whether they attend HBCUs or HWCUs.

9

## Students' learning opportunities

- African American students at HBCUs are more actively and deeply involved in the academic community than are their counterparts at HWCUs

- One of my studies also indicates less satisfying and more difficult academic experiences among African American female students at HWCUs.

10

## Reasons for "no difference"

- There are obvious compensating factors, and the two types of institution contribute to student learning in different ways: HWCUs provide more visible monetary resources and prestige, while HBCUs offer greater humane support and deeper involvement.
- In addition, the discriminatory climate at HWCUs might have eased since the desegregation movement.

11

## Conclusion

- HBCUs appear to be more cost-effective in achieving their mission of educating black students. How they manage to produce the same level of outcomes as HWCUs in spite of poorer academic and financial resources needs to be investigated further.
- Given that HBCUs are significantly underfunded relative to HWCUs, the findings of my studies and of other reports lend support to the proposition that HBCUs contribute significantly to higher education in this country and merit strong support.

12

**Table 1.**
**Comparing Student and Institutional Characteristics of HBCUs and HWCUs: Means, Standard Deviations, and Correlation Coefficients**

| | HWCU | | HBCU | | r with Black college |
|---|---|---|---|---|---|
| | Means | SD | Means | SD | |
| **Student characteristics** | | | | | |
| Degree completion | 0.63 | 0.48 | 0.55 | 0.50 | −0.08 |
| Respondent's early earnings | 4.23 | 1.76 | 3.97 | 1.65 | −0.08 |
| Parental income | 6.96 | 3.28 | 6.25 | 3.21 | −0.11 * |
| High school GPA | 5.55 | 1.67 | 4.56 | 1.69 | −0.29 * |
| College GPA [2] | 3.67 | 0.95 | 3.79 | 1.04 | 0.06 |
| **Institutional characteristics** | | | | | |
| *Global characteristics* | | | | | |
| Selectivity in admission | 1105.00 | 120.63 | 724.80 | 77.32 | −0.73 * |
| Total enrollment | 14486.94 | 10864.51 | 1823.40 | 1002.92 | −0.67 * |
| *Internal characteristics* | | | | | |
| Student-faculty ratio | 20.88 | 5.76 | 15.80 | 2.78 | −0.42 * |
| Percentage: research project with faculty | 22.96 | 7.23 | 31.89 | 5.77 | 0.50 * |
| Student-faculty interaction | 7.92 | 0.18 | 8.06 | 0.16 | 0.27 * |
| Average faculty salary | 42751.94 | 6579.04 | 26814.00 | 5470.87 | −0.68 * |
| Percentage: faculty with a Ph.D. | 84.53 | 6.19 | 62.80 | 12.20 | −0.68 * |
| Percentage: total instruction-related expenditure | 73.13 | 11.61 | 72.35 | 8.80 | −0.03 |
| Instruction-related expenditure per FTE student | 9246 | 5306 | 6506 | 1208 | −0.25 * |
| Percentage: students transferring | 12.25 | 7.48 | 15.71 | 6.41 | 0.22 * |
| Percentage: undergraduates receiving aid | 53.53 | 13.18 | 84.00 | 11.10 | 0.67 * |

Note: [1] The correlation coefficients are Spearman's rho; * $p < 0.05$ (two-tailed)
[2] The information of the variable was much reduced, and the available samples of college GPA for mean, SD, and correlation analyses were only 280.
Most of the comparative data were drawn from Kim (2002) and Kim and Conrad (2006).

## REFERENCES


W. Allen, "The color of success: African-American college student outcomes at Predominantly White and Historically Black public colleges and universities," *Harvard Educational Review*, vol. 62(1) (1992), pp. 26– 44.

A. W. Austin, What Matters in College: Four Critical Years Revisited. (San Francisco: Jossey-Bass, 1993).

J. Baratz, and M. Ficklen, *Participation of recent Black college graduates in the labor market and in graduate education* (Research Report, Education Policy Research Institute). (Princeton, NJ: Educational Testing Service, 1983).

L. Bohr, E. T. Pascarella, A. Nora, and P. T. Terenzini, "Do Black students learn more at Historically Black or Predominantly White Colleges?" *Journal of College Student Development*, vol. 36(1) (1995), pp. 75–85. J. Centra, R. Linn, and M. Parry, "Academic growth in predominantly Negro and predominantly White colleges," *American Educational Research Journal*, vol. 7(1) (1970), pp. 83–98.

J. M. Constantine, "The effect of attending Historically Black Colleges and Universities on future wages of Black students." *Industrial and Labor Relations Review*, vol. 48(3) (1995), pp. 531–46.

P. Cross, and H. Astin, "Factors influencing Black students' persistence in college," in *Black Students in Higher Education,* ed. G. Thomas (Westport, CN: Greenwood Press, 1981), pp. 76–90.).

R. G. Ehrenberg, and D. S. Rothstein, "Do historically Black institutions of higher education confer unique advantages on Black students: An initial analysis" (Cambridge, MA: National Bureau of Economic Research, 1993). Unpublished paper.

D. H. Jackson, "Attracting and retaining African-American faculty at HBCUs," *Education*, vol. 123(1) (2002), pp. 181–185.

M. M. Kim and C.F. Conrad, "The Impact of Historically Black Colleges and Universities on the Academic Success of African-American Students," *Research in Higher Education,* vol. 47 (4) (2006), pp. 399–427. M.M. Kim, "Historically Black vs. White Institutions: Academic Development among Black Students," *The Review ofHigher Education,* vol. 25(4) (2002), pp. 385–407.

E. Pascarella and P. Terenzini, *How College Affects Students* (San Francisco: Jossey-Bass, 1991).

E. Pascarella, M. Edison, A. Nora, L. S. Hagedorn, and P. Terenzini, "Additional evidence on the cognitive effects of college racial composition: A research note," *Journal of College Student Development*, vol. 37(5) (1996), pp. 494–501.

E. Pascarella, J. Smart, C. Ethington, and M. Nettles, M, "The influence of college on self-concept: A consideration of race and gender differences," *American Educational Research Journal*, vol. 24(1) (1987), pp. 49–77.

E. Pascarella, J. Smart, and J. Stoecker, "College race and the early status attainment of Black students," *The Journal ofHigher Education*, vol. 60(1) (1989), pp. 82–107.

S. Raudenbush and A. Bryk, A. *Hierarchical Linear Models: Applications and Data Analysis Methods* (2nd ed.) (Thousand Oaks, CA: Sage Publication, 2002).

L. M. Solnick, "Black college attendance and job success of Black college graduates," *Economics ofEducation Review*, vol. 9(2) (1990), pp. 135–148.

C. V. Willie, and R. R. Edmonds, *Black Colleges in America: Challenge, Development, Survival* (New York: Teachers College Press, 1978).

## Speaker Biographies

**LOUIS W. SULLIVAN** is the founding Dean and first President of Morehouse School of Medicine (MSM) in Atlanta, GA. With the exception of his tenure as Secretary of the U.S. Department of Health and Human Services (HHS) from 1989 to 1993, Dr. Sullivan was President of MSM for more than two decades. On July 1, 2002, he left the presidency, but continues to assist in national fund-raising activities on behalf of the school and is an Adjunct Professor of Medicine.

A native of Atlanta, Dr. Sullivan graduated magna cum laude from Morehouse College, and earned his medical degree, cum laude, from Boston University School of Medicine. He is certified in internal medicine and hematology.

Dr. Sullivan became the founding Dean and Director of the Medical Education Program at Morehouse College in 1975. He left MSM in 1989 to join President George H.W. Bush's Cabinet as the Secretary of the U.S. Department of Health and Human Services (HHS). Dr. Sullivan's tenure (47 months) stands as one of the longest of any HHS secretary in U.S. history.

As head of HHS, Dr. Sullivan managed the federal agency responsible for the major health, welfare, food and drug safety, medical research, and income security programs that serve the American people. During his time as Secretary, HHS undertook the most extensive food labeling reform in the country's history, initiated the Human Genome Project, implemented major drug approval process reforms, including "parallel track," and combined 65 separate programs for children and families (and a supporting staff of 2,000) under a new, unified operating division, the Administration for Children and Families.

In January 1993, he returned to MSM and resumed the office of president.

Dr. Sullivan is Chairman of the Board of the National Health Museum in Washington, D.C. and is also Chairman of the Sullivan Commission on Diversity in the Healthcare Workforce (funded by the Kellogg Foundation). He also serves as Chair of the President's Board of Advisors on Historically Black Colleges and Universities, a member of the U. S. Secretary of Education's Commission on the Future of Higher Education, and is Co-Chair of the President's Advisory Council on HIV and AIDS.

A member of numerous medical organizations, including the American Medical Association and the National Medical Association, Dr. Sullivan was the founding President of the Association of Minority Health Professions Schools. He is a former member of the Joint Committee on Health Policy of the Association of American Universities and the national Association of Land Grant Colleges and Universities.

**EARL S. RICHARDSON** was appointed the 11th president of Morgan State University on November 1, 1984, after serving eight months as interim president. He took the helm of the University at one of the most critical moments in its history, a period of declining enrollment, fiscal uncertainty, a deteriorating physical plant and considerable anxiety both within and outside the institution about its future. The versatility and grace with which he has guided the University and the course on which he has led it since that time have distinguished his presidency as one of the great periods in the history of Morgan.

President Richardson joined the Morgan family as a 41-year old scholar holding the Bachelor of Arts Degree in social science from the University of Maryland Eastern Shore (formerly Maryland State College) and both the Master of Science Degree and Doctor of Education Degree in education administration from the University of Pennsylvania. He came to Morgan with a wealth of teaching and administrative experience extending over a 19 year period: Assistant to the President, University of Maryland System and Executive Assistant to the Chancellor, Director of Career Planning and Placement and acting Director of Admissions and Registration at the University of Maryland Eastern Shore. This administrative experience was buttressed by solution-oriented research as fellows for the Ford and Kellogg Foundations and the publication of several articles on the implications of proposals to merge historically Black colleges with white institutions and on inter-institutional cooperation in higher education, all combining to make him a recognized authority on problems in higher education relative to racial autonomy, desegregation and integration.

Since becoming president of Morgan State University, Dr. Richardson has fashioned an all-encompassing strategy for strengthening the University: a growing inventory of graduate and undergraduate programs aimed at new technology and the global community of the 21$^{st}$ century; enhancement of the credentials and scholarly profile of the faculty; the establishment of national centers of excellence in a number of critical fields; and a variety of initiatives to increase minority representation in the professions and to ease the transition of students from high school and community colleges into four-year universities; a rapidly growing and greatly refurbished physical plant; burgeoning increases of over 75 percent in student enrollment, coupled with dramatic increases in the quality of students; sweeping improvements in external funding and government, corporate and foundation support. As a result of his leadership, Morgan has experienced noteworthy growth and improvement over the last two decades and is in the midst of the Renaissance that it is now celebrating.

Dr. Richardson is active in civic and community organizations. He is a member of the Boards of Directors of LifeBridge, Good Samaritan Hospital, the Council for Higher Education on Accreditation and the Goldseeker Foundation; serves on the advisory board or honorary board of the Maryland Coalition Foundation, Baltimore Neighborhoods, Inc. and the Quality Education for Minorities Network; and sits on the Board of Trustees of the Education Testing Service.

Dr. Richardson is married to the former Sheila Bunting of Oklahoma City, Oklahoma, and they have a son, Eric.

**JAMIE P. MERISOTIS** is the founding President of the Institute for Higher Education Policy. Established in 1993 in Washington, DC, the Institute is regarded as one of the world's premier research and policy organizations concerned with higher education policy development. As the Institute's President, Mr. Merisotis has worked extensively on nearly every aspect of the Institute's work. He is recognized as a leading authority on college and university financing, particularly student financial aid, and has published major studies and reports on topics ranging from higher education ranking systems to technology-based learning. Mr. Merisotis also manages the Institute's growing global portfolio, working to further educational opportunity and access primarily in nations in transition, such as in Southern Africa and the former Soviet Union.

A champion of the idea that higher education reaps rich rewards for both society and individuals, Mr. Merisotis has focused much of his work on improving access to higher education for low-income, minority, and other underrepresented populations. This commitment to equality of opportunity was a major factor in the establishment of the Alliance for Equity in Higher Education, an unprecedented coalition of national associations that represent more than 350 minority-serving colleges and universities, including Historically Black Colleges and Universities, American Indian Tribal Colleges, and Hispanic-Serving Institutions. Founded in 1999, the Alliance serves as a leading voice for the interests of minority-serving institutions (MSIs) and has become a model of collaboration, unity, and innovation among communities of color. Mr. Merisotis serves as the coordinator and facilitator of the Alliance and is director of the Kellogg MSI Leadership Fellows Program, a year-long, intensive program designed to train the next generation of presidents at MSIs.

Prior to founding the Institute, Mr. Merisotis served as Executive Director of the National Commission on Responsibilities for Financing Postsecondary Education, a bipartisan commission appointed by the President and the Congressional leadership. He is a member of the Board of Trustees of Bates College in Lewiston, Maine, and is Trustee and Chair-Elect of Scholarship America, the nation's largest private sector scholarship and educational support organization. Mr. Merisotis is the recipient of numerous awards and honors, including the 2002 Robert P. Huff Golden Quill Award from the National Association of Student Financial Aid Administrators and the 2001 Community College Government Relations Award presented by the American Association of Community Colleges and Association of Community College Trustees. He was a 2005 finalist for the Brock International Prize in Education, and in 1998 was named one of the top young leaders (under the age of 45) in American higher education by *Change* magazine.

**RAYMOND C. PIERCE** was appointed Dean of North Carolina Central University School of Law in July 2005. Prior to this appointment Pierce had a successful career with the national law firm of Baker & Hostetler LLP. As a partner in the firm's office in Cleveland, Ohio, Dean Pierce's legal practice concentrated in business transactions and public policy. In addition, he served state governments in higher education and pension investment-related matters and was also a member of the law firm's Federal Policy Group based in Washington D.C. where he assisted clients with government-related issues. Prior to joining Baker & Hostetler, Dean Pierce was a candidate for mayor of Cleveland.

From 1993 to 2000, Pierce served as President Bill Clinton's Deputy Assistant Secretary for Civil Rights at the U.S. Department of Education. While deputy assistant secretary, Dean Pierce managed the enforcement of federal civil rights laws in education and the development of federal civil rights education policy. Dean Pierce led the development of the Clinton Administration's federal education and civil rights policy in response to the 1992 U.S. Supreme Court decision in *Ayers v Fordice,* 505 U.S. 717 (1992), which addressed equal protection in higher education opportunities for African-American students and the impact of state policies on Historically Black Colleges and Universities (HBCUs). Following the publishing of the *Fordice Notice*, Dean Pierce directed the Department of Education's national higher education desegregation docket and negotiated resolution agreements with six states resulting in increased funding and enhanced educational programming for HBCUs. Through these efforts, Dean Pierce was the commencement speaker at graduation ceremonies for five HBCUs on behalf of the Clinton Administration. He also served on committees with the White House Domestic Policy Council where he assisted in the development of federal urban economic development initiatives. Under Dean Pierce's leadership, his agency received three Vice-President Awards for improving operations in the federal government. During his service with the federal government, Dean Pierce was appointed by the U.S. Secretary of Education to serve on three Senior Executive Service Appointment Boards. Dean Pierce has lectured extensively on the subject of civil rights and education and served on numerous panels and task forces.

Prior to joining the Clinton Administration, Dean Pierce was employed by the LTV Corporation from 1986–993, beginning as a Labor Relations Counselor and later serving as an attorney in that company's Corporate Law Department. While in the Corporate Law Department at LTV, Pierce was a commercial transactions attorney primarily involved with the transportation section for railroads, waterways and trucking.

Dean Pierce began his legal career in 1984 as a civil rights attorney with the John W. Walker Law Firm in Little Rock, Arkansas where he concentrated on class action employment discrimination cases. He is also a past Chairman of the Education Committee of the Cleveland Branch of the NAACP.

Dean Pierce is the recipient of numerous awards including the Thurgood Marshall Scholarship Fund National Leadership Award, an NAACP National Service Award and the Case Western Reserve University School of Law Distinguished Graduate Award.

Dean Pierce received his law degree from Case Western Research University School of Law in 1983 and his bachelor's degree in 1980 from Syracuse University, where he was admitted into the Sigma Tau Delta English Honorary Society. He is admitted to practice in Ohio and Arkansas.

**MIKYONG MINSUN KIM** (PhD, UCLA) is an Associate Professor of higher education and the Director of the Virginia Campus Higher Education Administration Doctoral Program of the George Washington University. Former posts include faculty positions at the University of Arizona - Tucson and the University of Missouri - Columbia. She also served as a grant panelist and consultant for the National Science Foundation.

She has been actively engaged in contributing to the field of higher education. Her teaching and research interests include college impact, comparative higher education, finance, equity and opportunity issues, higher education policy, institutional analysis, special colleges for special populations, organizational theory, and bridging K-12, higher education and work. While she encompasses a wide range of interest areas, she has dedicated a great deal of her focus on the impact and effectiveness of Historically Black Colleges and Universities on African-American students. Currently she has an article scheduled to be released in June 2006, *Research in Higher Education*, entitled, "The Impact of Historically Black Colleges and Universities on the Academic Success of African-American Students." Contact: kimmi@gwu.edu, (703) 726-3771, (202) 973-1510.

# Commissioner Statements

## Statement of Commissioner Abigail Thernstrom

HBCUs raise an obvious question: Halfa century after Brown v. Board of Education, should we still support institutions of higher education that are overwhelmingly black?

In 1992, the Supreme Court held in U.S. v. Fordice that racially identifiable institutions of higher education in Mississippi were a holdover from the state's Jim Crow past. The plaintiffs wanted more state funding for historically black schools (which are often in dire financial straits), but the court turned them down, seeing the ghost of "separate but equal."

The district court, which then had to fashion a remedy for the dual system, proposed eliminating all but one of the traditionally black colleges in the state. If only one of the HBCUs in Mississippi remained, that would solve the problem of racially separate campuses, since it would force almost all black students to attend schools that were majority white. The proposal prompted the NAACP, in the summer of 1994, to organize a civil-rights march to save Mississippi Valley State University. In a strange turn of events, the preservation of an allegedly "segregated" institution became a black cause. "It's important. . . for blacks to hang onto something and call it their own," one student explained to the Chronicle of Higher Education.

Almost all of the HBCUs, which now number approximately 100, were created at a time when Southern blacks were excluded from other schools. (Only four of the HBCUs are outside the South.) They turned out doctors, lawyers, ministers and politicians. W.E.B. Du Bois graduated from Fisk in 1888, Thurgood Marshall from Lincoln (Chester County, Pa.) in 1930, and Martin Luther King Jr. from Morehouse (Atlanta) in 1948. As recently as 1960, these schools were the choice of almost two-thirds of black college students.

They are no longer exclusively for blacks. But they are racially identifiable—much more so than most of the urban public schools that are routinely denounced as disastrously "segregated." Well over 80 percent of the students, and more than six out of 10 faculty members, are African-American. These institutions are defined by a commitment to a black identity that, for many students, remains meaningful. Morehouse, for instance, proudly advertises itself as "The Best College in the Nation for Educating African-Americans."

But overwhelmingly white institutions are also eager for a significant black presence. If all educational doors are so wide open to black students and professors, should we really hope that these financially struggling schools survive? At the height of the civil-rights movement, some thoughtful observers expressed grave doubts. Research by Christopher Jencks and David Riesman concluded that these "by-products of the Southern caste system" were usually an "ill-financed, ill-staffed caricature of white higher education." The "great majority [stood] near the end of the academic procession in terms of student aptitudes, faculty competence, and intellectual ferment." And it seemed self-evident that the demise of Jim Crow would make these schools even weaker by exposing them to new competition. Their best students and teachers would have choices denied earlier generations of African-Americans, and few would choose even the better HBCUs like Howard over Harvard. In fact, while the proportion of black students educated in HBCUs has fallen sharply with a significant rise in the total number of blacks attending college, in absolute numbers enrollment in these schools has grown substantially.

Perhaps they continue to thrive because they do an excellent job of making sure their students actually get a diploma. Getting into college is not the great problem for blacks in higher education today; staying and graduating is. The dropout rate at the HBCUs is high, but considerably lower than at the typical majority-white school. Although only a 10th of all African-Americans attend HBCUs, they award over a fifth of all bachelor's degrees earned by blacks. And yet they do not attract students from unusually affluent and educated homes.

Admittedly, their academic standards may still be relatively low. They tend to take students, as one advocate put it, who arrive academically "rough around the edges." But the HBCUs have an advantage over even the selective traditionally white colleges: There is no mismatch between black student qualifications and the academic demands of the schools. No students are preferentially admitted -- given a break in a system of racial double standards. And once they enroll at an HBCU, they can feel free to major in more difficult subjects, knowing that they will not be unprepared for the coursework.

In fact, a remarkable 40% of all African-Americans with a bachelor's degree in the physical sciences, and 38% of those who majored in math or the biological sciences, attended HBCUs. Moreover, among blacks who earned a Ph.D. in the late 1990s, 31% had done their undergraduate work at HBCUs. Messrs. Jencks and Riesman, writing in the '60s, complained that the schools were "traditional," rather than "progressive" and "harked back to an earlier period of white academic history." But their academic conservatism may be the secret to their success.

An impressive study by the economists Roland Fryer and Michael Greenstone suggests that HBCUs are not what they used to be. In the 1970s, compared with black students at other institutions, students at HBCUs went on to jobs that paid significantly higher wages. By the 1990s that labor-market advantage had disappeared. There was no "brain drain" into better white schools; the measurable academic credentials of black students at HBCUs relative to those at largely white schools had not changed. So what happened?

It is possible that the Fryer and Greenstone finding may reflect the special historical circumstances of the 1970s, when employers first came under pressure to hire a more racially diverse work force and found it most efficient to comply with affirmative-action mandates by recruiting at schools in which most job applicants would be African-American. By the 1990s these pressures may have eased, or employers may have developed other strategies for dealing with them.

Whatever the true story (and the inadequacies of the data make a definitive judgment difficult), the evidence does not indicate that HBCUs are "the best" schools for African-Americans across the board, as some enthusiasts claim. But they do seem to meet a real need, serving their students well in important respects.

In a free society, many private and public institutions will have a distinctive demographic profile. Group clustering is not necessarily unhealthy; indeed, it's an inescapable feature of a multiethnic nation. No one worries that there are "too many" Jews at Yeshiva and Brandeis, "too many" Catholics at Notre Dame and Holy Cross, "too many" Mormons at Brigham Young. And so it should be with Howard, Fisk and Mississippi Valley State. That's what democratic pluralism means.

This comment rests heavily on an op-ed of mine, co-authored with Stephan Thernstrom. "Separation Anxiety," The Wall Street Journal, November 30, 2007, W13.

## Dissent of Commissioners Michael Yaki and Arlan D. Melendez

As has been necessary for us to do with regard to other recent Briefing Hearing Reports approved and issued by the Commission's conservative majority, we are compelled to voice objection to the manner in which this Report was prepared. Further, we are able to support neither its content nor any of its Findings and Recommendations.

This Report is shamefully tardy. The Briefing Hearing about which it seeks to illuminate the concerned public took place in May, 2006, over four full years prior to the majority's final approval of its Findings and Recommendations. The delay was interposed by Commissioners who wished, for reasons related to their own political purposes, to marry this Report to the Commission's Briefing Hearing Report entitled "Encouraging Minority Students to Pursue Science, Technology, Engineering and Math Careers" which is being published simultaneously with this Report. There is no valid reason that this HBCU Report could not have been published during the more than two full years between the Briefing Hearing upon which it is predicated and the STEM Briefing Hearing held in 2008.

Further, parts of this Report's conclusions rest upon stale data, some of which was introduced into the record long after the Briefing Hearing and was over a decade old at the time of its initial consideration. This data is now almost 15 years old,[244] nearly the same age as today's students whose post-secondary educational opportunities the majority wishes to impact via its "advice" in Recommendation #1.

We are further concerned that Finding #6 and Recommendation #1 are included merely to trumpet the notion that African-American students may suffer unduly due to "mismatch" between their academic abilities and the willingness of highly selective and competitive schools to admit them. This theory was not discussed at the Briefing Hearing, but rather was grafted onto the Report thanks to an extraordinary reopening of the record.

## Rebuttal by Commissioner Gail Heriot

***August 26, 2010***

Commissioners Yaki and Melendez call this report "shamefully tardy." "Shamefully" is a strong word. Although I would much prefer to curl up with a good book and a nice cup of tea this evening, it would be imprudent to let such an accusation go unanswered.

It was for good reason that this report was placed in the queue behind some of our other work. Chairman Reynolds, who unquestionably has both the authority and duty to set the agenda, made the judgment that it would get more readers and hence be more effective if it were revamped and released at the same time as our report entitled "Encouraging Minority Students to Pursue Science, Technology, Engineering and Math Careers." I understand that these two reports will indeed be published on the same day. That decision is not just something I support. I suggested it (or, more precisely, I suggested something similar to the course of action the Chairman ultimately took and happily supported his alternative course).

Revamping a draft report that otherwise could not have garnered majority support and releasing it in conjunction with a related report is not just appropriate conduct. It would have been inappropriate to issue the report in the condition that it came to us from the Commission staff. And it would have been inappropriate to be indifferent to the increased readership the report could have if released in conjunction with a related report. The Chairman's decision in no way prevented the Commission from accomplishing its business; it simply changed the order of that business.[245]

The objections of Commissioners Yaki and Melendez—and this is by no means the first time they have raised them--are perplexing. There is no requirement that the Commission release reports in the order that the briefings were held, just as there is no requirement that Congress legislate in the order that bills were introduced or in the order that legislative hearings were conducted. Indeed, there is no requirement that the Commission release a report following each briefing at all.[246]

I find it hard to believe that the Speaker of the House would do anything but laugh if her appointee, Commissioner Yaki, were to apply the same argument to Congress that he has been applying to the Commission. Of course, the Speaker and other Congressional leaders make judgments about which bills need to be revamped. They supplement records when appropriate. They take proposals in the order that they believe are best, all things considered, rather than in the order that bills were introduced or that legislative hearings were held. And yes, they consider the likely public response to Congressional activities in setting the agenda. If they didn't, they would be derelict in their duty.

Commissioners Yaki and Melendez do not appear to disagree with the Commission's majority that the initial draft of this report needed substantial re-direction.[247] Nor have they argued that the report will have not more readers now that it is being released in conjunction with "Encouraging Minority Students to Pursue Science, Technology, Engineering and Math Careers." Indeed, the possibility of that increased readership appears to be what has inflamed them.

Commissioners Yaki and Melendez have all but admitted that they would prefer that readers not know that a much larger proportion of African-American graduates of HBCUs major in science and engineering than African-American graduates of mainstream colleges and universities. But it is simply true that HBCUs do more than their share of producing African-American scientists and engineers. These commissioners also appear to prefer that readers not learn the reason that some scholars have found for this: As a result of race-conscious admissions policies, African-American students at mainstream colleges and universities tend to have academic credentials that put them towards the bottom of their college classes. These scholars have found that an aspiring science or engineering major who attends a school where her entering academic credentials put her in the middle or the top of her class is more likely to follow through with that ambition than an otherwise identical student attending a more elite school where *those same credentials* place her towards the bottom of the class. As a result, African-American students receiving a preference at an elite mainstream institution are at a disadvantage when it comes to majoring in science and engineering. This is not a problem at HBCUs and hence explains their superior performance in this area.[248]

At least one HBCU faculty member–Professor Walter Pattillo, Jr. of North Carolina Central University–intuitively grasped the mismatch problem even before scholars were able to demonstrate its existence empirically. As then-chairman of the biology department, he vented his frustrations to *Science* in 1992: "The way we see it, the majority schools are wasting large numbers of good students. They have black students with admission statistics [that are] very high, tops. But these students wind up majoring in sociology or recreation or get wiped out altogether."[249]

Unlike some deliberative bodies, the Commission's procedures permit members to alter the agenda or to require measures be brought up for a vote even if it is not the Chairman's pleasure. But to do that, of course, they must have a majority of the Commission behind them. Commissioners Yaki and Melendez did not. When a motion was made to require that this report be released before its companion report could be prepared, it was defeated. There is nothing "shameful" or even particularly unusual about the Commission's decision.

## End Notes

[1] During the course of this briefing our panelists and commissioners used a wide variety of terms to describe a college or university that is not an HBCU. These terms included *traditionally white institutions, historically white colleges and universities,* and *predominantly white schools,* among others. For clarity in this report we used the term *non-HBCU* except when directly referencing a phrase, sentence or quotation from the prepared statements or the transcript where the panelist or author used a different but synonymous term. In these cases we used the term or phrase used by the panelist or author.

Terms such as *traditionally white institutions* and *historically white colleges and universities* should not be construed to mean that such institutions are racially segregated or that blacks or other minorities are discouraged from attending those institutions. In fact, many institutions which our panelists describe using terms such as *historically white college or university* are quite racially diverse.

[2] Roland G. Fryer, Jr., and Michael Greenstone, "The Causes and Consequences of Attending Historically Black Colleges and Universities," National Bureau of Economic Research Working Paper No. 13036, April 2007.

[3] Rogers Elliott, A. Christopher Strenta, Russell Adair, Michael Matier, and Jannah Scott, "The Role of Ethnicity in Choosing and Leaving Science in Highly Selective Institutions, " *Research in Higher Education*, vol. 37, No. 6. (1996).

[4] Rogers Elliott, A. Christopher Strenta, Russell Adair, Michael Matier, and Jannah Scott (1996), "The Role of Ethnicity in Choosing and Leaving Science in Highly Selective Institutions," *Research in Higher Education*, Vol. 37, No. 6, p. 684, citing E. Culotta (1992), "Black colleges cultivate scientists," *Science* 258: 1216–1218 and D. H. Thurgood and J. E. Clark (1995), *Summary Report 1993: Doctorate Recipients from United States Universities*. Washington, DC: The National Academy Press.

[5] "The National Campus Compact is a coalition of more than 950 institutions committed to the civic purposes of higher education." USCCR briefing transcript, May 5, 2006, p. 27.

[6] Rogers Elliott, A. Christopher Strenta, Russell Adair, Michael Matier, and Jannah Scott (1996), "The Role of Ethnicity in Choosing and Leaving Science in Highly Selective Institutions," *Research in Higher Education*, Vol. 37, No. 6, p. 690, citing G. H. Dunteman, J. Wisenbaker, and M. E. Taylor (1979), *Race and Sex Differences in College Science Program Participation*. Research Triangle Park, N.C.: Research Triangle Institute and J. Oakes (1990), "Opportunities, Achievement, and Choice: Women and Minority Students in Science and Mathematics," *Review ofResearch in Higher Education* 16:153–222.

[7] Rogers Elliott, A. Christopher Strenta, Russell Adair, Michael Matier, and Jannah Scott (1996), "The Role of Ethnicity in Choosing and Leaving Science in Highly Selective Institutions," *Research in Higher Education*, Vol. 37, No. 6, p.690.

[8] *Adams v. Richardson*, 351F. Supp. 636 (1972) and *Adams v. Califano*, 430 F. Supp. 118 (D.D.C. 1977).

[9] USCCR briefing transcript, May 5, 2006, p. 42. Unnecessary program duplication refers to those instances in which broadly similar high demand programs at an historically black college or university (HBCU) are offered at a geographically proximate TWI. *See*, for example, MARYLAND'S REPORT AND THE PARTNERSHIP AGREEMENT BETWEEN THE STATE OF MARYLAND AND THE U. S. DEPARTMENT OF EDUCATION, OFFICE FOR CIVIL RIGHTS, <http://www.mhec.state.md.us/higherEd/ocrplan/index.asp> (last accessed Jan. 26, 2007) The HBCU and TWI in question may be public or private. Genevieve Segura senior education policy analyst, Maryland Higher Education Commission, telephone interview, Jan. 26, 2007.

[10] Rogers Elliott, A. Christopher Strenta, Russell Adair, Michael Matier, and Jannah Scott (1996), "The Role of Ethnicity in Choosing and Leaving Science in Highly Selective Institutions," *Research in Higher Education*, Vol. 37, No. 6, p. 682, citing J. Oakes (1990), "Opportunities, achievement, and choice: Women and minority students in science and mathematics," *Review ofResearch in Higher Education* 16:153–222; L. Suter (ed.) (1993), *Indicators of Science and Mathematics—1993*. Washington, D.C.: U. S. Government Printing Office (NSF 93–95); P. E. White (1992), *Women and Minorities in Science and Engineering: An Update*. Washington, D.C.: National Science Foundation.

[11] "Factors influencing Black students' persistence in college," by P. Cross, P. and H. Astin, in G. Thomas (Ed.). Black Students in Higher Education, 1981, pp. 76–90. Westport, CN: Greenwood Press; "Do historically Black institutions of higher education confer unique advantages on Black students: An initial analysis," by. R. G. Ehrenberg, and D. S. Rothstein, Cambridge, MA: National Bureau of Economic Research, 1993, unpublished paper; "The influence of college on self-concept: A consideration of race and gender differences," by E. Pascarella, J. Smart, C. Ethington, and M. Nettles, in American Educational Research Journal, 24 (1), 1987, pp. 49-77. Kim written statement, pp. 46, 56–57.

[12] Rogers Elliott, A. Christopher Strenta, Russell Adair, Michael Matier, and Jannah Scott (1996), "The Role of Ethnicity in Choosing and Leaving Science in Highly Selective Institutions," *Research in Higher Education*, Vol. 37, No. 6, p. 700, citing *The College Handbook*, for example, 1988, or any recent edition; *Barron's Profiles ofAmerican Colleges*, for example, 1988, or any recent edition; E. Culotta (1992), "Black colleges cultivate scientists," *Science* 258: 1216–1218 and T. R. Phillips (1991), *ABET/EXXON Minority Engineering Student Achievement Profile.*

[13] Rogers Elliott, A. Christopher Strenta, Russell Adair, Michael Matier, and Jannah Scott (1996), "The Role of Ethnicity in Choosing and Leaving Science in Highly Selective Institutions," *Research in Higher Education*, Vol. 37, No. 6, p. 700, citing D. H. Thurgood and J. E. Clark (1995, Table 5), *Summary Report 1993: Doctorate Recipients from United States Universities*. Washington, DC: The National Academy Press.

[14] Rogers Elliott, A. Christopher Strenta, Russell Adair, Michael Matier, and Jannah Scott (1996), "The Role of Ethnicity in Choosing and Leaving Science in Highly Selective Institutions," *Research in Higher Education*, Vol. 37, No. 6, pp. 701–702.

[15] John A. Moore, Note: Are State-Supported Historically Black Colleges and Universities Justifiable after Fordice?—A Higher Education Dilemma," 27 Fla. St. U. L. Rev. 547, 549 (2000).

[16] Earl S. Richardson, president, Morgan State University written statement to the U.S. Commission on Civil Rights, Washington, DC, May 5, 2006, p. 33 (hereafter cited as Richardson written statement).

[17] John A. Moore, Note: Are State-Supported Historically Black Colleges and Universities Justifiable after Fordice?—A Higher Education Dilemma," 27 Fla. St. U. L. Rev. 547, 549 (2000).

[18] Debra A. Harley, "Desegregation at HBCUs: Removing Barriers and Implementing Strategies." The Negro Education Review, Vol. 52, No. 4, October 2001, p. 151.

[19] United States Department of Education, President's Board of Advisors on Historically Black Colleges and Universities, "White House Initiative on Historically Black Colleges and Universities," <http://www.ed.gov/ about/inits/list/whhbcu/edlite-index.html> (last accessed Jan. 19, 2007).

[20] Samuel L. Meyers, "What is a Black College?" NAFEO Inroads, (February /March, 1987), p. 2.

[21] PBS, From Swastika to Jim Crow, "Historically Black Colleges and Universities," <http://www.pbs.org/itvs/fromswastikatojimcrow/Black colleges.html> (last accessed October 16, 2005).

[22] PBS, From Swastika to Jim Crow, "Historically Black Colleges and Universities," <http://www.pbs.org/itvs/fromswastikatojimcrow/Black colleges.html> (last accessed October 16, 2005).

[23] Samuel L. Meyers "What is a Black College?" NAFEO Inroads, (February./March., 1987), p. 2.

[24] John A. Moore, Note: Are State-Supported Historically Black Colleges and Universities Justifiable after Fordice?—A Higher Education Dilemma," 27 Florida State University Law Review 547, 549–550 (2000).

[25] John A. Moore, Note: Are State-Supported Historically Black Colleges and Universities Justifiable after Fordice?—A Higher Education Dilemma," 27 Fla. St. U. L. Rev. 547, 550 (2000); PBS, From Swastika to Jim Crow, "Historically Black College and Universities," <http://www.pbs.org /itvs/fromswastikatojimcrow/Blackcolleges_2.html> (last accessed October 16, 2005).

[26] John A. Moore, Note: Are State-Supported Historically Black Colleges and Universities Justifiable after Fordice?—A Higher Education Dilemma," 27 Fla. St. U. L. Rev. 547, 550 (2000).

[27] John A. Moore, Note: Are State-Supported Historically Black Colleges and Universities Justifiable after Fordice?—A Higher Education Dilemma," 27 Fla. St. U. L. Rev. 547, 550 (2000); PBS, From Swastika to Jim Crow, "Historically Black College and Universities," <http://www.pbs.org/itvs/ fromswastikatojimcrow/Blackcolleges_2.html> (last accessed October 16, 2005); Leonard H. O. Spearman, Governance in the Nation's Historically Black Colleges and Universities,, 2005, p. 7.

[28] National Center for Educational Statistics, Historically Black Colleges and Universities, 1976–2001, September 2004, p.2, Table 1.

[29] Roland G. Fryer, Jr., and Michael Greenstone, “The Causes and Consequences of Attending Historically Black Colleges and Universities,” National Bureau of Economic Research Working Paper No. 13036, April 2007, p.1.

[30] 505 U.S. 717 (1992).

[31] United Negro College Fund, <www.uncf.org/print/index.asp?prlD=174) (last accessed Nov. 23rd, 2007).

[32] The Executive Summary and the Summary of the Proceedings are based on the briefing held on May 5, 2006. A transcript of the briefing is available on the Commission’s Web site <http://www.usccr.gov> and by request from the Publications Office, U.S. Commission on Civil Rights, 624 Ninth Street, NW, Room 600, Washington, D.C. 20425. The telephone number and e-mail address of the Publications Office are (202) 376-8128 and publications@usccr.gov, respectively.

[33] The acronyms TWI, HWI, HWCU, and non-HBCU are used at various points in this report to reflect the differing terminology used by our panelists. For purposes of this report these terms mean the same thing. When referring globally or collectively to the panelist and or commissioner comments this report uses the term nonHBCU. In addition, note that it would be interesting to assess the impact of attending an Historically Black College or University (HBCU) on white students, but this is beyond the scope of the Commission’s discussion.

[34] Roland G. Fryer, Jr., and Michael Greenstone, “The Causes and Consequences of Attending Historically Black Colleges and Universities,” National Bureau of Economic Research Working Paper No. 13036, April 2007, p. 18.

[35] Roland G. Fryer, Jr., and Michael Greenstone, “The Causes and Consequences of Attending Historically Black Colleges and Universities,” National Bureau of Economic Research Working Paper No. 13036, April 2007, pp. 18–25.

[36] Roland G. Fryer, Jr., and Michael Greenstone, “The Causes and Consequences of Attending Historically Black Colleges and Universities,” National Bureau of Economic Research Working Paper No. 13036, April 2007, Appendix, Table 6, p. 46.

[37] Rogers Elliott, A. Christopher Strenta, Russell Adair, Michael Matier, and Jannah Scott, “The Role of Ethnicity in Choosing and Leaving Science in Highly Selective Institutions, ” Research in Higher Education, vol. 37, No. 6. (1996), p. 682.

[38] The article did not name the Ivy-league schools. See Rogers Elliott, A. Christopher Strenta, Russell Adair, Michael Matier, and Jannah Scott, “The Role of Ethnicity in Choosing and Leaving Science in Highly Selective Institutions, ” Research in Higher Education, vol. 37, No. 6. (1996), p. 682

[39] Rogers Elliott, A. Christopher Strenta, Russell Adair, Michael Matier, and Jannah Scott, “The Role of Ethnicity in Choosing and Leaving Science in Highly Selective Institutions, ” Research in Higher Education, vol. 37, No. 6. (1996), p. 682.

[40] Rogers Elliott, A. Christopher Strenta, Russell Adair, Michael Matier, and Jannah Scott, “The Role of Ethnicity in Choosing and Leaving Science in Highly Selective Institutions, ” Research in Higher Education, vol. 37, No. 6. (1996), p. 695.

[41] Using a composite predictor of academic achievement, the authors found that, in statistical terms, white and Asian American students enjoy an advantage amounting to 1.75 standard deviations over black students. Note that in this study the averages of students’ combined SAT scores were: 1325 for white matriculants; 1160 for blacks; 1345 for Asians; and 1219 for Hispanics. Rogers Elliott, A. Christopher Strenta, Russell Adair,

Michael Matier, and Jannah Scott, “The Role of Ethnicity in Choosing and Leaving Science in Highly Selective Institutions, ”Research in Higher Education, vol. 37, No. 6. (1996), pp. 682 and 700.

[42] Rogers Elliott, A. Christopher Strenta, Russell Adair, Michael Matier, and Jannah Scott, “The Role of Ethnicity in Choosing and Leaving Science in Highly Selective Institutions, ”Research in Higher Education, vol. 37, No. 6. (1996), p. 684.

[43] The test, originally known as the Scholastic Aptitude Test, was later renamed the Scholastic Assessment Test, and now is simply called the SAT. See Fair Test: The National Center for Fair and Open Testing, “The SAT Questions and Answers,” <http://www.fairtest.org/facts /satfact.htm.>, (last accessed Aug. 31, 2006).

[44] Rogers Elliott, A. Christopher Strenta, Russell Adair, Michael Matier, and Jannah Scott, “The Role of Ethnicity in Choosing and Leaving Science in Highly Selective Institutions, ”Research in Higher Education, vol. 37, No. 6. (1996), pp. 684, 700.

[45] Rogers Elliott, A. Christopher Strenta, Russell Adair, Michael Matier, and Jannah Scott, “The Role of Ethnicity in Choosing and Leaving Science in Highly Selective Institutions, ”Research in Higher Education, vol. 37, No. 6. (1996), p. 700.

[46] Rogers Elliott, A. Christopher Strenta, Russell Adair, Michael Matier, and Jannah Scott, “The Role of Ethnicity in Choosing and Leaving Science in Highly Selective Institutions, ”Research in Higher Education, vol. 37, No. 6. (1996), pp. 700–701.

[47] U.S. Commission on Civil Rights, briefing on the educational effectiveness of historically black colleges and universities, Washington, DC, May 5, 2006 transcript (hereafter cited as USCCR briefing transcript, May 5, 2006), testimony of Louis W. Sullivan, p. 11.

[48] USCCR briefing transcript, May 5, 2006, p. 11.

[49] USCCR briefing transcript, May 5, 2006, p. 11.

[50] USCCR briefing transcript, May 5, 2006, p. 12.

[51] USCCR briefing transcript, May 5, 2006, p. 12.

[52] USCCR briefing transcript, May 5, 2006, pp 12–13.45

[53] USCCR briefing transcript, May 5, 2006, p. 13.

[54] USCCR briefing transcript, May 5, 2006, pp. 13–14.

[55] USCCR briefing transcript, May 5, 2006, p. 14.

[56] USCCR briefing transcript, May 5, 2006, pp. 14–15.

[57] Richardson written statement, p. 31.

[58] USCCR briefing transcript, May 5, 2006, pp. 16–17.

[59] USCCR briefing transcript, May 5, 2006, pp. 16–17; Richardson written statement, p. 31.

[60] Richardson written statement, p. 31.

[61] USCCR briefing transcript, May 5, 2006, pp. 17–18; Richardson written statement p. 31.

[62] USCCR briefing transcript, May 5, 2006, p. 18.

[63] Richardson written statement, p. 31.

[64] Richardson written statement, p. 31.

[65] Richardson written statement, p. 31.

[66] Richardson written statement, p. 31. From kindergarten through the 12th grade, poor funding of black schools resulted in closure and caused black students to enroll in predominantly white schools that were once closed to them. See Richardson written statement, p. 31.

[67] Richardson written statement, p. 31.

[68] Richardson written statement, pp. 31–32.

[69] USCCR briefing transcript, pp. 19–20.
[70] USCCR briefing transcript, p. 20.
[71] USCCR briefing transcript, p. 20.
[72] USCCR briefing transcript, p. 20.
[73] Richardson written statement, p. 32
[74] USCCR briefing transcript, May 5, 2006, pp. 21–22; Richardson written statement, p. 34
[75] Richardson written statement, p. 34.
[76] Richardson written statement, p. 32.
[77] USCCR briefing transcript, May 5, 2006, p. 22; Richardson written statement, p. 32
[78] Richardson written statement, pp. 33–34.
[79] See note 29.
[80] Richardson written statement, p. 33.
[81] Richardson written statement, p. 33.
[82] Richardson written statement, pp.33–34.
[83] USCCR briefing transcript, May 5, 2006, p. 22; Richardson written statement, p. 34.
[84] USCCR briefing transcript, May 5, 2006, p. 24
[85] USCCR briefing transcript, May 5, 2006, p. 25. Other panelists, for example, Dr. Sullivan also noted this factor. USCCR briefing transcript, May 5, 2006, pp. 14, 56–57. The National Survey of Student Engagement (NSSE), administered by the Indiana University Center for Post Secondary Research, documents the extent to which institutions are involved in educational practices or activities that result in preferred learning and personal development outcomes. NSSE collects information from students attending colleges and universities nationwide about their participation in institutional programs and activities designed for "learning and personal development." The findings estimate "how undergraduates spend their time and what they gain from attending college." NSSE began in 2000 and since then close to 1,000 colleges and universities have participated. Jamie P. Merisotis, president of the Institute for Higher Education Policy, written statement to the U.S. Commission on Civil Rights (hereafter cited as Merisotis written statement), Washington, DC, May 5, 2006, p. 25.
[86] USCCR briefing transcript, May 5, 2006, p. 26. Other panelists, for example Dr. Kim also noted this factor. USCCR briefing transcript, May 5, 2006, p. 44.
[87] USCCR briefing transcript, May 5, 2006, p. 26.
[88] USCCR briefing transcript, May 5, 2006, p. 26.
[89] Merisotis written statement, p. 37.
[90] Merisotis written statement, p. 38.
[91] "The National Campus Compact is a coalition of more than 950 institutions committed to the civic purposes of higher education." USCCR briefing transcript, May 5, 2006, p. 27.
[92] USCCR briefing transcript, May 5, 2006, p. 28.
[93] USCCR briefing transcript, May 5, 2006, p. 28.
[94] USCCR briefing transcript, May 5, 2006, pp. 28–29.
[95] The federally funded TRIO program historically has provided educational opportunities for low-income, first-generation college attendees, and disabled persons. The TRIO program includes Upward Bound, Talent Search and Student Support Services, and is authorized under the Higher Education Act. TRIO serves students from the pre-college to the pre-graduate levels. USCCR briefing transcript, May 5, 2006, p. 29.
[96] USCCR briefing transcript, May 5, 2006, p. 29.

[97] USCCR briefing transcript, May 5, 2006, p. 30.
[98] USCCR briefing transcript, May 5, 2006, p. 30.
[99] USCCR briefing transcript, May 5, 2006, p. 31.
[100] USCCR briefing transcript, May 5, 2006, p. 31.
[101] USCCR briefing transcript, May 5, 2006, pp. 31–32.
[102] USCCR briefing transcript, May 5, 2006, pp. 32–33.
[103] USCCR briefing transcript, May 5, 2006, p. 33.
[104] USCCR briefing transcript, May 5, 2006, p. 33.
[105] USCCR briefing transcript, May 5, 2006, p. 33. Brown v. Board ofEducation, 347 U. S. 483 (1954).
[106] Raymond C. Perce written statement to the U.S. Commission on Civil Rights, Washington, DC, May 5, 2006, p.42 (hereafter cited as Pierce written statement).
[107] Adams v. Richardson, 351F. Supp. 636 (1972) and Adams v. Califano, 430 F. Supp. 118 (D.D.C. 1977).
[108] USCCR briefing transcript, May 5, 2006, pp. 33–34; Pierce written statement, p. 42.
[109] USCCR briefing transcript, May 5, 2006, p. 34.
[110] Pierce written statement, p. 42.
[111] Pierce written statement, p. 42.
[112] Pierce written statement, p. 42.
[113] USCCR briefing transcript, May 5, 2006, p. 34; Pierce written statement, p. 43.
[114] USCCR briefing transcript, May 5, 2006, pp. 34–35; Pierce written statement, p. 43.
[115] USCCR briefing transcript, May 5, 2006, pp. 35–36; Pierce written statement, pp. 42–43.
[116] In a check list approach, such questions may be asked, "did the states enhance public HBCUs and implement affirmative action procedures at TWIs?" Pierce written statement, p. 43.
[117] USCCR briefing transcript, May 5, 2006, p. 37; Pierce written statement, p. 43.
[118] Pierce written statement, p. 43.These six states were Florida, Kentucky, Maryland, Pennsylvania, Texas, and Virginia. USCCR briefing transcript, May 5, 2006, p. 38; Pierce written statement, p. 43.
[119] USCCR briefing transcript, May 5, 2006, p. 38; United States v. Fordice, 505 U. S. 717 (1992)
[120] USCCR briefing transcript, May 5, 2006, pp. 38–39.
[121] USCCR briefing transcript, May 5, 2006, pp. 39–40.
[122] USCCR briefing transcript, May 5, 2006, p. 40.
[123] USCCR briefing transcript, May 5, 2006, pp. 39–40.
[124] USCCR briefing transcript, May 5, 2006, p. 40; Pierce written statement, p. 41.
[125] USCCR briefing transcript, May 5, 2006, p. 41; Pierce written statement, p. 44.
[126] USCCR briefing transcript, May 5, 2006, p. 42. Unnecessary program duplication refers to those instances in which broadly similar high demand programs at an historically black college or university (HBCU) are offered at a geographically proximate traditionally white institution (TWI). See, for example, MARYLAND'S REPORT AND THE PARTNERSHIP AGREEMENT BETWEEN THE STATE OF MARYLAND AND THE U. S. DEPARTMENT OF EDUCATION, OFFICE FOR CIVIL RIGHTS, <http://www.mhec.state.md.us/higherEd/ocrplan/index.asp> (last accessed Jan. 26, 2007). The HBCU and TWI in question may be public or private. Genevieve Segura

senior education policy analyst, Maryland Higher Education Commission, telephone interview, Jan. 26, 2007.

[127] USCCR briefing transcript, May 5, 2006, p. 42.

[128] USCCR briefing transcript, May 5, 2006, p. 45; Kim written statement, p. 46.

[129] USCCR briefing transcript, May 5, 2006, p.43; Kim written statement, p. 45.

[130] Kim written statement, Washington, DC, May 5, 2006, p. 45; Mikyong Minsun Kim, PowerPoint document to the U.S. Commission on Civil Rights, Washington, DC, May 5, 2006, p. 49 (hereafter cited as Kim PowerPoint document).

[131] USCCR briefing transcript, May 5, 2006, p. 45; Kim written statement, p. 50.

[132] USCCR briefing transcript, May 5, 2006, p. 44; Kim written statement, p. 45.

[133] Kim written statement, p. 45.

[134] Kim written statement, p. 45.

[135] Rogers Elliott, A. Christopher Strenta, Russell Adair, Michael Matier, and Jannah Scott (1996), "The Role of Ethnicity in Choosing and Leaving Science in Highly Selective Institutions," Research in Higher Education, Vol. 37, No. 6, p. 682, citing J. Oakes (1990), "Opportunities, achievement, and choice: Women and minority students in science and mathematics," Review ofResearch in Higher Education 16:153–222; L. Suter (ed.) (1993), Indicators of Science and Mathematics—1993. Washington, D.C.: U. S. Government Printing Office (NSF 93–95); P.E. White (1992), Women and Minorities in Science and Engineering: An Update. Washington, D.C.: National Science Foundation.

[136] Roland G. Fryer, Jr., and Michael Greenstone, "The Causes and Consequences of Attending Historically Black Colleges and Universities," National Bureau of Economic Research Working Paper No. 13036, April 2007, p. 9.

[137] Kim written statement, p. 45.

[138] Kim written statement, pp. 45, 47; Kim PowerPoint document, Table 1, p.58.

[139] The national six-year graduation rate for first-time, full-time degree-seeking African-American freshmen attending HBCUs who graduated in academic year 2005/2006 is 37 percent. The comparable figure for African-American freshmen attending TWIs is 42.7 percent. Andrew Mary, statistician, National Center for Education Statistics made several special data runs for the Commission drawing on data from the Integrated Postsecondary Education Data System (IPEDS). See Andrew Mary e-mail, October 4, 2007. Dr. Kim's graduation statistics are clearly based on different parameters. What is consistent between her graduation statistics and those retrieved from IPEDS data is that African-Americans who attended TWIs have higher graduation rates than their peers who went to HBCUs. Yet the difference is much smaller than might otherwise be expected when considering the difference in academic credentials alone between blacks who attend HBCUs and TWIs.

[140] USCCR briefing transcript, May 5, 2006, p. 44.

[141] USCCR briefing transcript, May 5, 2006, p. 45; Kim written statement, p. 46.

[142] Kim written statement, pp. 49–50; USCCR briefing transcript, May 5, 2006, p. 50.

[143] Dr. Kim used hierarchical linear and non-linear modeling, which, according to her reviews of statistical procedures, has "well-established advantages over standard regression techniques for evaluating the effectiveness of schools and colleges and handling multi-level nested data sets." USCCR briefing transcript, May 5, 2006, p. 45; Kim written statement, p. 46. She explained that she controlled for selectivity, enrollment, SAT scores, grades, high school grade point average, pre-test measures, and parental income

and level of education. In other words, the HBCU students and African-American TWI students in her sample are comparable with regard to these factors. For example, parental income level would be generally similar for both groups of students. Dr. Kim indicated that she did not control for regional effects, that is, the HBCUs and TWIs in her sample are not found in the same type of locale. They might be located in urban, rural, or suburban areas. Kim written statement, pp. 49–50.

144 USCCR briefing transcript, May 5, 2006, pp. 45–46; Kim written statement, p. 46.

145 "Do Black students learn more at Historically Black or Predominantly White Colleges?" by L. Bohr, E.T. Pascarella, A. Nora, A., and P. T. Terenzini in Journal of College Student Development, 36(1), 1995, pp. 75–85; "Academic growth in predominantly Negro and predominantly White colleges," by J. Centra, R. Linn, and M. Parry, in American Educational Research Journal, 7(1), 1970, 83-98; and "Additional evidence on the cognitive effects of college racial composition: A research note," by E. T. Pascarella, M. Edison, A. Nora, L. S. Hagedorn, and P. Terenzini, P. in Journal of College Student Development, 37(5), 1996, 494-501. Kim written statement, pp. 46, 56–57.

146 "Factors influencing Black students' persistence in college," by P. Cross, P. and H. Astin, in G. Thomas (Ed.). Black Students in Higher Education, 1981, pp. 76–90. Westport, CN: Greenwood Press; "Do historically Black institutions of higher education confer unique advantages on Black students: An initial analysis," by R. G. Ehrenberg, and D. S. Rothstein, Cambridge, MA: National Bureau of Economic Research, 1993, unpublished paper; "The influence of college on self-concept: A consideration of race and gender differences," by E. Pascarella, J. Smart, C. Ethington, and M. Nettles, in American Educational Research Journal, 24 (1), 1987, pp. 49-77. Kim written statement, pp. 46, 56–57.

147 Dr. Kim's finding of no significant difference was reflected in E. Pascarella, J. Smart, and J. Stoecker "College, race and the early status attainment of Black students," The Journal ofHigher Education, 1989 60(1), pp. 82-107 and R. G. Ehrenberg, and D. S. Rothstein, "Do historically Black institutions of higher education confer unique advantages on Black students: An initial analysis," Cambridge, MA: National Bureau of Economic Research, 1993, unpublished paper.

Research contrary to Dr. Kim's finding of no significant difference includes Roland Fryer and Michael Greenstone, "The Cause and Consequences of Attending Historically Black Colleges and Universities," NBER Working Paper No. 13036, April 2007, pp. 1–53; and J. Baratz, and M. Ficklen, "Participation of recent Black college graduates in the labor market and in graduate education," Research Report, Education Policy Research Institute, Princeton, NJ: Educational Testing Service, 1983. Her finding was also contrary to J. M. Constantine, "The effect of attending Historically Black Colleges and Universities on future wages of Black students," Industrial and Labor Relations Review, 48(3), 1995, pp. 531–46. Kim written statement, pp. 46, 56–57.

148 USCCR briefing transcript, May 5, 2006, p. 46; Kim written statement, p. 46.

149 Roland G. Fryer, Jr., and Michael Greenstone, "The Causes and Consequences of Attending Historically Black Colleges and Universities," National Bureau of Economic Research Working Paper No. 13036, April 2007, pp. 25–26.

150 Roland G. Fryer, Jr., and Michael Greenstone, "The Causes and Consequences of Attending Historically Black Colleges and Universities," National Bureau of Economic Research Working Paper No. 13036, April 2007, p. 2.

[151] USCCR briefing transcript, May 5, 2006, pp. 45–46; Kim written statement, p. 4.
[152] USCCR briefing transcript, May 5, 2006, pp. 46–47.
[153] USCCR briefing transcript, May 5, 2006, p. 47.
[154] USCCR briefing transcript, May 5, 2006, p. 47; Kim written statement, p. 47.
[155] USCCR briefing transcript, May 5, 2006, p. 47; Kim written statement, p. 47.
[156] USCCR briefing transcript, May 5, 2006, pp. 48–49.
[157] USCCR briefing transcript, May 5, 2006, pp. 49–50.
[158] USCCR briefing transcript, May 5, 2006, pp. 48–49, 51.
[159] USCCR briefing transcript, May 5, 2006, pp. 50–51.
[160] USCCR briefing transcript, May 5, 2006, p. 52.
[161] USCCR briefing transcript, May 5, 2006, pp. 51–52.
[162] USCCR briefing transcript, May 5, 2006, p. 52.

During his review of this draft report, Dean Pierce elaborated on his statements: "North Carolina Central University School of Law has a comparatively higher attrition rate amongst the HBCU Law Schools. NCCU School of Law also has the highest bar passage rate amongst the HBCU Law Schools and one of the highest bar passage rates for African-Americans amongst all law schools in the nation. NCCU School of Law continues its mission as a school of opportunity and admits many students with comparatively lower Law School Admission Test (LSAT) scores and lower grade point averages (GPAs). Grading is tough at the Law School for first year students. Students not achieving a minimum 2.0 GPA at the end of the end of the first year are academically dismissed." See Dean Pierce's comments on the Summary of Proceedings, Nov. 30, 2007, p. 2.

[163] USCCR briefing transcript, May 5, 2006, pp. 52–53.
[164] USCCR briefing transcript, May 5, 2006, p 53.
[165] USCCR briefing transcript, May 5, 2006, p. 54.
[166] USCCR briefing transcript, May 5, 2006, pp. 57–58
[167] USCCR briefing transcript, May 5, 2006, pp. 57–58.
[168] USCCR briefing transcript, May 5, 2006, pp. 58–59.
[169] USCCR briefing transcript, May 5, 2006, pp. 77.
[170] USCCR briefing transcript, May 5, 2006, pp. 77.
[171] USCCR briefing transcript, May 5, 2006, p. 78.
[172] USCCR briefing transcript, May 5, 2006, pp. 78.
[173] USCCR briefing transcript, May 5, 2006, pp. 81–82.
[174] USCCR briefing transcript, May 5, 2006, p. 82.
[175] Kim written statement, Table 1, p. 50.
[176] USCCR briefing transcript, May 5, 2006, p. 79.
[177] USCCR briefing transcript, May 5, 2006, p. 79.
[178] USCCR briefing transcript, May 5, 2006, pp. 82–83.
[179] USCCR briefing transcript, May 5, 2006, p. 83.
[180] USCCR briefing transcript, May 5, 2006, pp. 54–55. According to Dr. Kim, African-American degree completion rate at HBCUs was 55 percent compared to 63 percent at HWCUs. USCCR briefing transcript, May 5, 2006, p. 45. At first glance, Chairman Reynolds and Dr. Kim's statements seem contradictory. Note, however, that Chairman Reynolds was referring to the attrition of poorly prepared students at TWIs. Dr. Kim, on

the other hand, addressed national-level degree completion rates for African-American students at both types of institutions regardless of their academic preparation level.

[181] USCCR briefing transcript, May 5, 2006, p. 55. [182]USCCR briefing transcript, May 5, 2006, pp. 56–57.

[183] Rogers Elliott, A. Christopher Strenta, Russell Adair, Michael Matier, and Jannah Scott (1996), "The Role of Ethnicity in Choosing and Leaving Science in Highly Selective Institutions," Research in Higher Education, Vol. 37, No. 6, pp. 701–702.

[184] Rogers Elliott, A. Christopher Strenta, Russell Adair, Michael Matier, and Jannah Scott (1996), "The Role of Ethnicity in Choosing and Leaving Science in Highly Selective Institutions," Research in Higher Education, Vol. 37, No. 6, p.684.

[185] USCCR briefing transcript, May 5, 2006, pp. 59–61.

[186] USCCR briefing transcript, May 5, 2006, p. 61.

[187] USCCR briefing transcript, May 5, 2006, pp. 62–63.

[188] USCCR briefing transcript, May 5, 2006, pp. 63–64.

[189] USCCR briefing transcript, May 5, 2006, pp. 64–65.

[190] USCCR briefing transcript, May 5, 2006, pp. 65–67.

[191] USCCR briefing transcript, May 5, 2006, p. 67.

[192] USCCR briefing transcript, May 5, 2006, pp. 67–68.

[193] The reference here is to the Carnegie Classification of institutions. For example, disparity in faculty salary still prevails even when comparing, for example, HBCU teaching institutions to TWI teaching institutions.

[194] USCCR briefing transcript, May 5, 2006, pp. 69–70.

[195] Vice Chair Thernstrom noted that Harvard faculty Gary Orfield is a "leading spokesman for integrated schools," USCCR briefing transcript, May 5, 2006, p. 70.

[196] USCCR briefing transcript, May 5, 2006, p. 70.

[197] USCCR briefing transcript, May 5, 2006, p. 71.

[198] USCCR briefing transcript, May 5, 2006, pp. 71, 89. United States v. Fordice, 505 U. S. 717 (1992).

[199] USCCR briefing transcript, May 5, 2006, pp. 72, 90.

[200] USCCR briefing transcript, May 5, 2006, pp. 90–91.

[201] USCCR briefing transcript, May 5, 2006, p. 73.

[202] USCCR briefing transcript, May 5, 2006, p. 73.

[203] USCCR briefing transcript, May 5, 2006, p. 73.

[204] USCCR briefing transcript, May 5, 2006, p. 74

[205] USCCR briefing transcript, May 5, 2006, p. 74–75.

[206] USCCR briefing transcript, May 5, 2006, pp. 75–76.

[207] USCCR briefing transcript, May 5, 2006, p. 75.

[208] USCCR briefing transcript, May 5, 2006, pp. 72–73.

[209] USCCR briefing transcript, May 5, 2006, pp. 83–84.

[210] USCCR briefing transcript, May 5, 2006, pp. 84–85.

[211] USCCR briefing transcript, May 5, 2006, p. 85.

[212] USCCR briefing transcript, May 5, 2006, p. 111.

[213] USCCR briefing transcript, May 5, 2006, p. 111.

[214] USCCR briefing transcript, May 5, 2006, p. 85.

[215] USCCR briefing transcript, May 5, 2006, pp. 86, 87.

[216] USCCR briefing transcript, May 5, 2006, pp. 87–89.

[217] USCCR briefing transcript, May 5, 2006, pp. 91–92.
[218] USCCR briefing transcript, May 5, 2006, pp. 92–93
[219] USCCR briefing transcript, May 5, 2006, pp. 93–95.
[220] USCCR briefing transcript, May 5, 2006, pp.,94–96
[221] USCCR briefing transcript, May 5, 2006, p. 99.
[222] USCCR briefing transcript, May 5, 2006, p. 95.
[223] USCCR briefing transcript, May 5, 2006, pp. 96–97
[224] USCCR briefing transcript, May 5, 2006, pp. 96–97.
[225] USCCR briefing transcript, May 5, 2006, pp. 97–98.
[226] USCCR briefing transcript, May 5, 2006, pp. 93, 98.
[227] USCCR briefing transcript, May 5, 2006, pp. 98–99.
[228] USCCR briefing transcript, May 5, 2006, pp. 99–100.
[229] USCCR briefing transcript, May 5, 2006, p 103.
[230] USCCR briefing transcript, May 5, 2006, pp. 102–104.
[231] USCCR briefing transcript, May 5, 2006, p. 80.
[232] USCCR briefing transcript, May 5, 2006, pp. 80–81.
[233] USCCR briefing transcript, May 5, 2006, p. 110.
[234] USCCR briefing transcript, May 5, 2006, p. 110.
[235] USCCR briefing transcript, May 5, 2006, p. 110.
[236] USCCR briefing transcript, May 5, 2006, p. 111–112.
[237] USCCR briefing transcript, May 5, 2006, pp. 112–113.
[238] USCCR briefing transcript, May 5, 2006, p. 114
[239] USCCR briefing transcript, May 5, 2006, p. 114.
[240] USCCR briefing transcript, May 5, 2006, p. 110.
[241] USCCR briefing transcript, May 5, 2006, p. 112.
[242] USCCR briefing transcript, May 5, 2006, p. 113.
[243] USCCR briefing transcript, May 5, 2006, pp. 113–114.
[244] Elliot, Rogers; Strenta, A. Christopher; Adair, Russell; Matier, Michael; and Scott, Jannah, "The Role of Ethnicity in Choosing and Leaving Science in Highly Selective Institutions," Research in Higher Education, Vol. 37, No. 6 (1996), pp. 681-709.
[245] In contrast, Commissioner Yaki's practice of walking out of meetings when doing so will defeat a quorum has indeed prevented the Commission from accomplishing its business on many occasions. As Commissioner Yaki once put it, "if you can't get your gang of five together, that's not my problem." See, e.g., Commission Business Meeting of September 11, 2009 (quorum defeated by Commissioner Yaki); Commission Business Meeting of July 11, 2008 (quorum defeated by Commissioner Yaki), Commission Business Meeting of August 16, 2006 (quorum defeated by Commissioner Yaki); Commission Telephone Meeting of October 30, 2009(quorum defeated by Commissioner Yaki); Commission Telephone Meeting of October 24, 2009 (quorum defeated by Commissioner Yaki). Commissioner Melendez has sometimes joined him in this behavior.

The quoted statement was from the Commission Telephone Meeting of October 30, 2009. At that conference, Commission Yaki suddenly announced "Commissioner Yaki is leaving the conference" directly after Commissioner Taylor was called away to attend to his legal practice. Tr. at 37. The remaining four commissioners remained on the line for an informational meeting on one of its projects for a few minutes, but were unable to take any votes, given the lack of a quorum. After the discussion ended and the court reporter retired, Commissioner Yaki mysteriously re-appeared on the call and, mistakenly believing the other commissioners had left the call, began to speak to staff members. It was likely that he had never hung up, since the tone that sounds when someone enters a call had never sounded. When asked about his conduct, he became irate. His remarks were recorded by a special assistant shortly after hearing them.

On November 11, 2009, he made similar statements on the record, declaring that "should you [i.e. Chairman Reynolds] choose to move forward with your agenda, which I disagree with, I presume that you will have the foresight to ensure that the five members of your coalition are present in order to effectuate that vote." "I will not be a party," he said, "to contributing to your ability to get things done." Transcript 8-9.

And indeed he has not been. On other occasions, Commissioner Yaki has mysteriously joined commission telephone meetings only after it has become apparent that the Commission has a quorum. See, e.g., Commission Telephone Meeting of September 3, 2009. Unlike most members of the Commission, Commissioner Yaki is never willing to state that he will be able to make one of the Commission's regularly scheduled monthly telephone conferences, always reserving for himself the maneuvering room necessary to defeat a quorum if the opportunity should present itself. This has led to the cancelling of meetings, because it is clear that a quorum will not be available without Commissioner Yaki's cooperation and that cooperation will not be forthcoming. See, e.g., Commission Telephone Meeting of August 27, 2010.

His temper tantrums, which we all have become accustomed to, don't help either. See, e.g., Commission Business Meeting of May 15, 2009 (Tr. at 80). Commissioner Heriot: "Let the record reflect that Commissioner Yaki has been screaming, yelling, and is now storming out of the room." Commissioner Yaki's reaction was to scream that he instead of "screaming" he was "raising his voice in utter disbelief"). See also Commission Business Meeting of December 4, 2009 (Transcript at 116).

I would have more sympathy for Commissioner Yaki's obstreperous conduct (1) if he did not then turn around and complain that our reports are "shamefully tardy" and (2) if he abandoned only those meetings in which some momentous decision is being made with which he disagrees. In fact, however, Commissioner Yaki only rarely passes up a chance to defeat a quorum, even when the business on the Commission's agenda is quite mundane.

[246] Although Commissioners Yaki and Melendez have to no standing to complain about the speed at which the Commission gets its reports completed, see supra at n. 1, if an outsider to the Commission were to remark that our briefing reports sometimes take longer to publish than is optimal, I would be inclined to agree. Part of the problem is the peculiar procedure we employ. Our briefings generally consist of a panel of experts who "brief" the Commission on a particular issue in civil rights. Witnesses are urged to submit their comments in writing prior to the briefing. At the briefing itself, they are asked to summarize their remarks orally, and those oral statements are recorded along with a question and answer period by a court reporter. The body of our briefing report consists of a summary of the oral summary and of the question and answer period drafted by a staff member. This "double summary" is make-work and ordinary takes about a year to produce. For many reports, simply assembling the written statements of the witnesses and attaching to transcript of the briefing as an appendix would require one tenth the work by

the staff. Moreover, such a procedure would also result in a more readable report. It would also allow greater time for the Commissioner's Statements and Rebuttals, which tends to be more useful than the "double summary," and could be more useful still if Commissioners (who usually work full-time at other jobs) had more than thirty days to produce them. Commissioner Taylor has suggested this streamlined approach in the past and I certainly support it.

247 Although the original briefing took prior to my tenure, I was on the Commission when Commission staff members produced a draft report a year later. The draft featured two recommendations for the President, Congress and the American people:

that "federal and state governments should vigorously support and appropriately fund HBCUs" and that "federal and state governments ... should take steps to avoid unnecessary duplication of HBCU academic programs at geographically proximate traditionally white institutions."

I was not inclined to vote for either of these recommendations and was confident that neither were a majority of my colleagues. This is not to say that the staff members who produced the draft were not well-intentioned. Staff members seldom get as much time as I would like to confer with Commission members about the kinds of recommendations Commission members want to make. In part this is a result of poorly-designed internal procedures. Sometimes staff members end up making guesses, and sometimes those guesses turn out to be incorrect. They learn. Ultimately, however, under the Commission's statute, it is for the Commission--and not the staff--to issue reports. As commissioners, we are responsible for ensuring their quality.

As for my reasons for opposing the specific recommendations in the draft report: Washington is full of people who would like to testify that they are doing a great job and that their work should be "vigorously supported." This is true whether the testimony is being taken by Congress, by a federal commission, or by one of the many advisory committees that meet in federal buildings every workday. Many of these witnesses are sincere. Some are even correct that they are doing a great job. But there was no reason for the Commission to weigh in on such a complex and multi-faceted issue. Indeed, it would be presumptuous for the Commission to instruct the federal and state government on such a matter. A lot goes into a decision to "vigorously support" and fund any particular institution. State and federal governments do not have unlimited funds. Money that is spent funding HBCUs is money that governments cannot spend on funding for public school lunch programs, agricultural subsidies, programs to protect forests or endangered species, highway or high speed rail construction, or national defense– to name just a few programs that commonly vie for federal or state dollars. We could not possibly address all of these competing priorities in any briefing, and many are not even within our jurisdiction. Recommending that state government give HBCUs monopoly positions in some of their programs struck me as narrower, which is good. But it also struck me as particularly unwise. Competition among state-funded colleges and universities is a good thing in some situations and a bad thing in others. The Commission was in no position to issue a blanket recommendation of this kind, given that it would almost certainly be misconstrued in the future.

The easiest and perhaps even the best thing to do at the time might have been to kill the report. The Commission has done this in the past when information-gathering briefings or hearings have, for one reason or another, not lead to any useful findings or recommendations. An example that comes to mind is the Commission Briefing on the School Districting Controversy in Omaha, Nebraska (September 8, 2006). This does not necessarily mean that the information gathering would be all for naught. In some of those cases, the Commission has published the briefing transcript, so that the sworn testimony contained in it can be available to policymakers, journalists, scholars and ordinary citizens who might be interested in it.

That, however, would have been an unhappy state of affairs. Although I could not support the recommendations in the draft and did not see any better recommendations that could be distilled from the report as it stood, I knew that there is plenty of evidence in the academic literature that HBCUs were training more than their proportionate share of the nation's African-American scientists, engineers and physicians. This much more limited claim I could support. I therefore told then-Staff Director Ken Marcus that while I did not favor a general declaration that HBCUs should be "vigorously supported," or that their programs should be given monopoly status by state higher education authorities, I would be very happy to give credit where credit is due to HBCUs—in their success in graduating a disproportionate number of the nation's African-American science, technology, engineering and mathematics majors. I furnished Mr. Marcus with one or two articles that confirmed this fact and told him that I believed others could be easily found. My understanding is that he discussed this with the Chairman Reynolds and that the Chairman agreed. In later discussions with Mr. Marcus, we talked about the desirability for further briefing from experts on the subject of encouraging minority students to pursue careers in science, technology, engineering and mathematics. Such a panel of experts was later assembled and the briefing held on September 12, 2008. Witnesses included Dr. Rogers Elliott, professor emeritus of psychology at Dartmouth College, Dr. Thomas Fortmann, a member of the Massachusetts Board of Elementary and Secondary Education and an electrical engineer, Dr. Richard Sander, professor of law at UCLA, Dr. Richard Tapia, professor of mathematics at Rice University and Robin Willner, Vice President of Global Community Initiatives at the IBM Corporation. President Hazel O'Leary of Fisk University, an HBCU, was originally scheduled to testify. She cancelled hours before the briefing (and not long after the list of witness participants was sent to the Commissioners and their special assistants).

I would have preferred to combine this report with the report on this second briefing. Chairman Reynolds preferred to issue them as separate reports on the same day. So be it. Such internal housekeeping issues are about as controversial as a paper clip to anyone except Commissioner Yaki and his allies. More time had to be spent by the Commission defending the Chairman's decision against the protests of Commissioner Yaki and his allies than can be justified.

[248] See Statement of Commissioner Gail Heriot, Encouraging Minority Students to Pursue Science, Technology, Engineering and Math Careers, United States Commission on Civil Rights Briefing Report (November __, 2010)(issued contemporaneously with this report). See also Rogers Elliott, A. Christopher Strenta, Russell Adair, Michael Matier & Jannah Scott, The Role of Ethnicity in Choosing and Leaving Science in Highly Selective Institutions, 37 Res. Higher Ed. 681, 692-93 (1996); Frederick Smyth & John J. McArdle, Ethnic and Gender Differences in Science Graduation at Selective Colleges with Implications for Admission Policy and College Choice, 45 Res. Higher Ed. 353

(2004); Richard Sander & Roger Bolus, Do Credentials Gaps in College Reduce the Number of Minority Science Graduates?, Working Paper (Draft July 2009).

Commissioners Yaki and Melendez have suggested that the evidence for mismatch in science and engineering in the Elliott study is old (and apparently in their view stale). It is not clear what they have to say about the Smyth & McArdle or the Sander & Bolus articles. It bears noting that most large empirical studies of the kind cited above are based on large databases. Often these databases are assembled not by the scholars themselves but by large research institutions attempting to facilitate research generally. The Smyth & McArdle study, for example, was performed on the College & Beyond database collected by the Andrew W. Mellon Foundation. These databases are massive undertakings. Just as the United States does not take a new census every year, these databases are not re-created every year. There is not a lot of evidence to suggest, however, that conditions have changed since then.

[249] Elizabeth Culotta, Black Colleges Cultivate Scientists, 258 Science 1216, 1218 (November 13, 1992).

In: Higher Education of People of Color
Editor: Dana G. Perry
ISBN: 978-1-63483-525-1


*Chapter 2*

# ENCOURAGING MINORITY STUDENTS TO PURSUE SCIENCE, TECHNOLOGY, ENGINEERING AND MATH CAREERS*

***U.S. Commission on Civil Rights***

**Letter of Transmittal**

The President
The President of the Senate The Speaker of the House

Sirs and Madam:

The United States Commission on Civil Rights ("Commission") is pleased to transmit this report, *Encouraging Minorities to Pursue Science, Technology, Engineering and Math Careers*. A panel of experts briefed the Commission on September 8, 2008, on possible reasons that minority students who begin college intending to major in science, technology, engineering or math ("STEM") leave these disciplines in disproportionate numbers before graduation. They also discussed possible ways to improve the retention of these students in STEM degree programs. Based on that briefing, the Commission developed the findings and reconunendations that are included in this report.

The Commission found that regardless of their racial or ethnic backgrounds, college freshmen show equally substantial degrees of interest in STEM careers. Despite similar levels of interest, the Commission found that black and Hispanic students are ultimately less likely to major in or obtain doctoral degrees in STEM disciplines than are whites and Asians. Data presented to the Commission indicated that racial and ethnic discrimination in college is not a substantial factor in these disproportionate STEM attrition rates. It found that academic mismatch one consequence of some schools' racially and ethnically preferential admissions policies—is an important reason for these disparities, however.

---

* This is an edited, reformatted and augmented version of a briefing report issued October 2010.

For example, data indicate that success in STEM majors depends both on a student's absolute entering academic credentials and his or her credentials relative to other students in his or her classes. When black and white students have the similar academic credentials black students are actually more likely than their white counterparts to obtain STEM degrees. Thus, the Commission ascribed the higher minority attrition from STEM programs to credentials gaps or "mismatch" stemming in part from racially preferential admissions policies.

The Commission recommended that selective colleges not admit any STEM student with a large deficit in academic credentials relative to its STEM median without fully informing that student of the potential impact of such deficit on that student. Such disclosure should include the school's record of graduating students with similar academic credentials in STEM majors. Similarly, the Commission urged high school guidance counselors to advise students about the impact of large deficits in academic credentials on success in a particular college's STEM program. It further noted that well-designed academic support programs can sometimes help students with modest deficits in credentials to succeed in STEM programs and advised schools to implement the best practices employed by such programs and make admitted students aware of their availability.

Part A, which consists of the body of this report, was approved on June 11, 2010 by Chairman Reynolds and Commissioners Gaziano, Heriot, Kirsanow and Taylor. Vice Chair Themstrom abstained, and Commissioner Yaki voted against. Vote tallies for each of the Commission's Endings and recommendations, which make up Part B of the report, are noted therein.

For the Commissioners,
Gerald A. Reynolds,
Chairman

## EXECUTIVE SUMMARY

The Commission held a briefing entitled, "Encouraging Minority Students to Pursue Science, Technology, Engineering and Math Careers." In particular, the Commission examined why minority college students who begin their college studies intending to major in science, technology, engineering or math (STEM) leave these disciplines in disproportionate numbers before graduation.

Experts appearing before the Commission were Professor Richard Sander of the University of California at Los Angeles (UCLA) Law School; Dr. Richard Tapia, Maxfield-Oshman Professor in Engineering at Rice University; Dr. Rogers Elliott, Professor Emeritus of Psychological and Brain Sciences at Dartmouth College; Dr. Thomas Fortmann, Massachusetts Board of Elementary and Secondary Education; and Ms. Robin Willner, Vice President of Global Community Initiatives at IBM Corporation.

Of particular interest to the Commission on this occasion was the "mismatch hypothesis." The mismatch hypothesis holds that students whose academic credentials are significantly different from the average student in the class may learn less than they would have learned in a class in which their academic credentials "matched" those of the average student. Mismatch may be positive or negative. Students who are positively mismatched – that is, their academic credentials significantly exceed those of their peers – may not be sufficiently challenged by the material. As a result, they may become bored or disengaged. Students who are negatively mismatched – that is, their academic credentials are significantly below those of their peers – may feel overwhelmed by the speed at which difficult material is being taught. They may get lost – even though they could have mastered the material had it only been taught at a slower rate.

Under this hypothesis, aggressive affirmative action or any admissions decision for largely nonacademic reasons can lead to negative mismatch for any student, including underrepresented minorities.[1] Well-meaning efforts to benefit these students can, if the mismatch hypothesis is correct, cause these students to drop out of STEM programs in disproportionate numbers. The result is fewer, not more, minority physicians, scientists, and engineers.

In this briefing, the term "mismatch" did not include the admission of students with small academic deficits who, with the kind of support offered by the colleges and universities they attend, would remain interested in STEM and able to successfully complete a program.

There was substantial agreement among the witnesses. None disputed the evidence that blacks and Hispanics are at least as likely to express interest in STEM majors as whites prior to attending college. None disputed the evidence that blacks and Hispanics abandon their STEM ambitions in greater proportions than do whites and Asians.

Two witnesses, Dr. Sander and Dr. Elliott, each presented an empirical study that supported the mismatch hypothesis at the undergraduate level. None of the other witnesses disputed either study. Indeed, to one extent or another, they all agreed on the peril that results when a minority student, usually unknowingly, accepts an offer of admission at a college or university at which he or she is mismatched. All the witnesses agreed that prospective students should be informed that his or her academic credentials are substantially below the average at a particular school. Students could then make an informed decision about which school to attend.

No one took the position that the elimination of mismatch in admissions would eliminate the disparities in average STEM credentials upon completion of high school between black and Hispanics students on the one hand and white and Asian students on the other. Mismatch is simply a piece of the puzzle at the college level. Rather than highlight mismatch, Dr. Fortmann emphasized the difficulty of attracting competent STEM teachers to K-12 schools, which he blamed for inadequate college preparation. Dr. Tapia conceded that the data on mismatch presented by Dr. Sander and Dr. Elliott "are entirely credible to me because they reflect what I have seen at Rice." While he stated that he did "not dispute the data," he took issue with what he perceived to be the conclusions they drew from it and advocated instead strong support and mentoring programs at the graduate level. He emphasized the need for "an equitable presence" of minorities at top research university graduate programs in STEM, where minority faculty members can serve as both role models and mentors. Ms. Willner discussed a report entitled, "Out Before the Game Begins: Hispanic Leaders Talk About What's Needed to Bring More Hispanic Youngsters into Science, Technology and Math Professions," sponsored by IBM. Based on this testimony, the Commission made the following findings and recommendations:

# Findings and Recommendations

## Findings

1) Science, technology, engineering, and mathematics (STEM) graduates are important to the U.S. economy because they enable the United States to maintain its preeminence in STEM fields. [Approved (7-0): Chairman Reynolds, Vice Chair Thernstrom, Commissioners Gaziano, Heriot, Kirsanow, Taylor, and Yaki voted in favor.]
2) Black and Hispanic high school seniors exhibit about the same degree of interest in pursuing STEM careers as white students (Asian students are still more interested). But despite these initially high levels of interest, black and Hispanic students are less likely to major in or obtain a doctoral degree in STEM disciplines than are whites and Asians. [Approved 4-2-1: Chairman Reynolds and Commissioners Gaziano, Heriot, and Kirsanow voted in favor; Vice Chair Thernstrom and Commissioner Yaki voted against; Commissioner Taylor abstained.]
3) Data presented to the Commission showed that success in a STEM major depends both on the student's absolute entering academic credentials and on the student's entering academic credentials relative to other students in the class. When a student is in a class in which his or her entry credentials are significantly different from the median student, the student is "mismatched" for that class. This mismatch causes a loss of learning, either because the positively mismatched student is not challenged by the material or because the negatively mismatched student feels overwhelmed by the speed at which the material is being taught. [Approved 4-1-2: Chairman Reynolds and Commissioners Gaziano, Heriot, and Kirsanow voted in favor; Commissioner Yaki voted against; Vice Chair Thernstrom and Commissioner Taylor abstained.]
4) Data presented to the Commission indicated that racial or ethnic discrimination in college was not a substantial factor in black and Hispanic college students' disproportionate attrition from STEM majors. The evidence showed that when black and white students have the same academic index scores, black students are more likely than white students to receive a STEM degree. [Approved 4-0-3: Chairman Reynolds and Commissioners Gaziano, Heriot and Kirsanow voted in favor; Vice Chair Thernstrom and Commissioners Taylor and Yaki abstained.]
5) The practice of racial and ethnic preferences is one way in which black and Hispanic students may be admitted to a college or university at which their entering academic credentials are significantly lower than those of their peers. When top tier colleges and universities use racial and ethnic preferences to recruit and admit minority students with academic credentials that are significantly below their median but match the median of lower tier colleges the resulting mismatch at the top tier institutions has a cascading effect through many lower tiers as each tier engages in racial and ethnic preferences to recruit and admit black and Hispanic students who do not match the mean in its respective tier. Although the consequence of this cascading mismatch is not the principal reason for the current disparities between blacks and Hispanics and whites and Asians in STEM (see Finding 3 regarding absolute

credentials), it is a significant reason. There are fewer black and Hispanic physicians, scientists and engineers today than there would have been if colleges and universities had not recruited and admitted black and Hispanic students with significantly lower academic credentials than their average student. [Approved 4-1-2: Chairman Reynolds and Commissioners Gaziano, Heriot, and Kirsanow voted in favor; Commissioner Yaki voted against; Vice Chair Thernstrom and Commissioner Taylor abstained.]

6) The high STEM-major attrition rate of students with credentials deficits indicates that many students and their parents may be unaware of the significance of mismatch for students' success in STEM fields because of the lack of institutional transparency. [Approved 4-1-2: Chairman Reynolds and Commissioners Gaziano, Heriot and Taylor voted in favor; Commissioner Yaki voted against; Vice Chair Thernstrom and Commissioner Kirsanow abstained.]
7) One panelist indicated that some graduate schools that have intensive support programs have been successful in ameliorating the effects of a moderate degree of mismatch. [Approved 4-1-2: Chairman Reynolds and Commissioners Gaziano, Heriot and Kirsanow voted in favor; Commissioner Yaki voted against; Vice Chair Thernstrom and Commissioner Taylor abstained.]

## Recommendations

1) A selective college or university should not admit any student with a large deficit in academic credentials relative to its median student without fully informing the student of the impact this deficit could have. Such deficits place students at a high risk of failure.
   [Approved 5-3: Chairman Reynolds and Commissioners Gaziano, Heriot, Kirsanow and Taylor voted in favor; Vice Chair Thernstrom and Commissioners Melendez and Yaki voted against.]
2) In addition to providing other appropriate support and advice to students interested in STEM majors and careers, high school guidance counselors should advise these students about the significant impact of large deficits in academic credentials on college performance. [Approved 5-3: Chairman Reynolds and Commissioners Gaziano, Heriot, Kirsanow and Taylor voted in favor; Vice Chair Thernstrom and Commissioners Melendez and Yaki voted against.]
3) Each individual student's right to decide which school to attend -- based on the best available evidence and with help from parents and advisors -- should be respected. To aid students with the decision-making process, schools with STEM programs should disclose to all admitted students their projected college grade point averages (and the range of error). Schools should also disclose to interested students the school's track record for graduating students with similar academic indices in STEM majors. [Approved 4-3-1: Chairman Reynolds and Commissioners Gaziano, Heriot and Taylor voted in favor; Vice Chair Thernstrom and Commissioners Melendez and Yaki voted against; Commissioner Kirsanow abstained.]
4) Well-designed academic support programs can sometimes help students with modest deficits in credentials to succeed in STEM programs. Schools should study and

implement the best practices employed by successful academic support programs. Schools should also routinely disclose information about academic support services to all admitted students. [Approved 5-3: Chairman Reynolds and Commissioners Gaziano, Heriot, Kirsanow and Taylor voted in favor; Vice Chair Thernstrom and Commissioners Melendez and Yaki voted against.]

5) K-12 schools should recruit qualified math and science teachers using, if necessary, pay adjustments and incentives. [Approved 7-0: Chairman Reynolds, Vice Chair Thernstrom, and Commissioners Gaziano, Heriot, Kirsanow, Taylor and Yaki voted in favor; Commissioner Melendez did not participate in the vote.]

# SUMMARY OF PROCEEDINGS

## Richard Sander

Professor Richard Sander[2] has both a Ph.D. in economics and a JD, and is a professor at UCLA Law School where he conducts empirical research on social policy. He began by discussing the different underrepresentation rates among major racial groups with respect to earning an undergraduate or doctoral degree in general education and in STEM disciplines.[3] His data analysis[4] indicated that black students were less likely than white students to earn a bachelor's degree of any kind or a doctorate, relative to their proportion of the general population. Professor Sander stated that underrepresentation in science is even more disproportionate: blacks are 36 percent as likely as whites to earn a bachelor's degree in science, 15 percent as likely to earn a doctorate in science, and 8 percent as likely to earn a non-biological science doctorate. He said that underrepresentation among Hispanics was similar to that of African-Americans with respect to undergraduate and graduate degrees, but that Asians were overrepresented in science relative to their population numbers. He added that within a science degree population, however, Hispanics are better represented relative to their concentrations in other disciplines.

Professor Sander set out four hypotheses to explain STEM underrepresentation among non-Asian minority students (slide 3):

- Black and Hispanic students are less interested in science than whites and Asians (Hypothesis 1);
- Black and Hispanic students have lower achievement levels and credentials by the time they finish high school, which affects their subsequent success rate (Hypothesis 2);
- Minority students have worse outcomes because of factors such as discrimination or inadequate support (Hypothesis 3); and
- Many capable minority students go into science but struggle or leave those disciplines because of mismatch. Professor Sander defined "mismatch" as any loss of learning that occurs because of a disparity between the credentials of a student and the median credentials of his classmates in a learning environment (Hypothesis 4).[5] Such disparities can be the result of admissions policies that emphasize affirmative action.

Professor Sander found evidence to support both Hypothesis 2 and Hypothesis 4. On the other hand, the evidence he presented contradicted Hypothesis 1 and Hypothesis 3.

Professor Sander presented data from several studies that showed black and Hispanic high school seniors are at least as interested in becoming science majors as whites. Based on this research, Sander argued that Hypothesis 1 (black and Hispanic students are simply less interested in science careers) is unsupported (slides 4, 5).

By contrast, Professor Sander found that his data supported Hypothesis 2 (black and Hispanic high school graduates have lower achievement levels and credentials that affect their STEM success rate) and was probably the major explanation for both their lower graduation rates in general and higher attrition rates in science in particular (slide 6).

Professor Sander explained that a widely-read book on affirmative action, *The Shape of the River,*[6] by William Bowen and Derek Bok, has somewhat obscured the relationship between entering academic credentials and academic success.

In that book, Bowen and Bok presented evidence that at elite schools, SAT scores are not a good predictor of the likelihood of graduation and that graduation rates are high from these schools regardless of SAT scores.[7] Sander stated that any conclusion that SAT scores don't matter to academic success or to graduation rates in general is fundamentally wrong and has not held up to further research.

To illustrate the relationship between entering academic credentials and academic success, he offered an analysis of data obtained from the University of Michigan on several thousand undergraduates who matriculated in 1999. This database made use of an "academic index" assigned to each student by the University of Michigan, which took into consideration both standardized test scores and high school record.

Grouping students into three cohorts based on whether they had received a large preference, a moderate preference, or no apparent preference, Sander found that black students who received a large preference had a four-year graduation rate of 21 percent, compared to a 73 percent four-year graduation rate for black students who received no preference (slide 10). A broadly similar pattern held for whites.

He also offered data from the National Longitudinal Survey of Freshmen, which showed that students, both black and white, with higher academic credentials have higher four-year graduation rates than those with lower academic credentials (slide 11).

Both sets of data showed also that blacks with high credentials had a higher four-year graduation rate than whites with similar credentials.

Professor Sander attempted to demonstrate that what is true for graduation rates also holds for concentration in the sciences: high credentials are important.

He observed that for the 1999 Michigan entering students, 5 percent of blacks who received substantial preferences ended up majoring in science or engineering, compared with 43 percent of blacks who received no preference (slide 9). His analysis revealed that whites displayed a similar pattern: 4 percent of those who received a large preference ended up majoring in science or engineering, versus 33 percent who received no preference. As with graduation, Dr. Sander stated that blacks with high academic credentials were more likely to major in science or engineering than were whites with the same credentials.

He determined that this not only supported Hypothesis 2 (credential levels have a significant impact in producing science graduates), but contradicted Hypothesis 3 (discrimination and inadequate academic support undermine black and Hispanic academic performance). If high-credential blacks are more likely to major in and graduate from science programs than high-credential whites at Michigan, then Dr. Sander did not think it likely that discrimination plays a meaningful role in STEM attrition.[8] He stated that the problem is that blacks (and to a lesser extent, Hispanics) are far more clustered in the lower-index/high-preference ranges at Michigan than are whites and Asians.

To examine Hypothesis 4 (the effect of "mismatch" on minority success in science), Dr. Sander turned to his analysis of a large data set that he obtained from the University of California, Office of the President (UCOP), that covered nearly a half-million UC students from 1992 through 2006 (slide 13). He stated that to test mismatch, one must compare students with similar credentials who are attending institutions with different degrees of admissions selectivity (slide 12).

He stated, for example, that a student with solid but not outstanding credentials might be close to the student median at UC Santa Barbara, but be able to attend UC Berkeley only with the benefit of a significant preference, and thus be potentially affected by mismatch (slide 13). He believed that the UCOP data set makes it possible to compare many thousands of similar students in different academic settings.

Dr. Sander said that his preliminary analysis of the UCOP data strongly supports his mismatch hypothesis (slides 14–19). When he compared students at Berkeley and UCLA whose credentials were substantially lower than the median with similar students attending other UC campuses, the students at less selective campuses had much higher probabilities of graduating with a science degree, often double the science graduation rate of such students at Berkeley and UCLA.

He said this held true whether or not one limited the analysis to students intending to major in science in college, or if one considered only underrepresented minority students or all students with potential mismatch at Berkeley or UCLA. It also held true at several different time periods.

Dr. Sander also compared students at UCLA and Berkeley whose credentials were equal to or stronger than their median classmates with similar students at other campuses. Since these students would not be mismatched at UCLA or Berkeley, mismatch theory would predict that they would not be at any disadvantage at those schools. Dr. Sander's data (slide 19) found that these students had higher science graduation rates at Berkeley and UCLA than the average rate for the median students at the other six UC campuses.

Professor Sander also reported on data from the Collegiate Learning Assessment study,[9] which attempts to measure student learning at nearly 200 American colleges (slide 20). He examined cross-section data on factors influencing whether a student was a math major, including as independent variables student credentials, "mismatch" (measured by the gap between a student's credentials and the average credentials of his peers at that college), race, and gender.

His analysis found that the degree of mismatch between student and school was the strongest (and very negative) predictor of seniors choosing to major in math; credentials were also strongly predictive, while neither sex nor race predicted pursuit of a math degree. He viewed this, too, as supporting hypotheses two and four (the role of credentials and mismatch) and as cutting against hypotheses one and three (student interest and discrimination).

(Note: Professor Sander's slides begin on the following page.)

Slide 1

## Does the "Mismatch Effect" Reduce the Ranks of Minority Scientists?

A Presentation to the
US Civil Rights Commission
September 12, 2008
Dr. Richard Sander, UCLA

Slide 2

## How Significant Is the Racial Gap in Science?

| Freq Rel to Pop | White | Black | Hispanic | Asian |
|---|---|---|---|---|
| Gen Pop | 100 | 100 | 100 | 100 |
| BA | 100 | 56 | 33 | 128 |
| Ph.D. | 100 | 43 | 21 | 130 |
| BA Science | 100 | 36 | 41 | 454 |
| Ph.D. Science | 100 | 15 | 26 | 703 |

Slide 3

## Four Possible Hypotheses

- Hypothesis 1: Black and Hispanic students are less interested in science than whites and Asians.
- Hypothesis 2: Lower achievement and credentials for Blacks and Hispanics by the time of high school graduation translates to lower success rates in science during and after college.
- Hypothesis 3: Minority college students studying science have worse outcomes than similar white and Asian students, because of factors like discrimination.
- Hypothesis 4: Many talented minorities interested in science are admitted through preferences into colleges where their credentials are much lower than those of their classmates. This "mismatch" causes them to struggle in science classes and either leave science or leave college.

Slide 4

## Hypothesis One: Are minorities less interested in science?

- The evidence is consistent and overwhelming that Black and Hispanic high school seniors are slightly *more* interested in a science career than are whites. (Asians are still more interested, by a substantial margin.)
- When credentials are taken into account, the higher level of Black and Hispanic interest is even more pronounced.

Slide 5

| Comparison Group | Proportion of students of each race intending to major in science | | | |
|---|---|---|---|---|
| | White | Black | Hispanic | Asian |
| University of California – Average | 34.7% | 37.5% | | 52.6% |
| University of California-2004-2006 | 37.3% | 40.5% | | 57.1% |
| HERI-CIRP data (Astin et al.) | 27.3% | 34.2% | 35.7% | 52.6% |
| Rogers Elliott data (4 elite institutions) | 41.4% | 44.2% | 44.0% | 55.0% |

Slide 6

## Hypothesis Two: Do achievement levels through high school shape later success in science?

- Yes, very much so.
- There's a mythology that's been fostered by many defenders of affirmative action that credentials have little or no impact upon success in college. What matters is not one's preparation going in, but which college one attends. Hence, preferences are crucial.
- This was a core proposition advanced by Bowen & Bok in *The Shape of the River*.

Slide 7

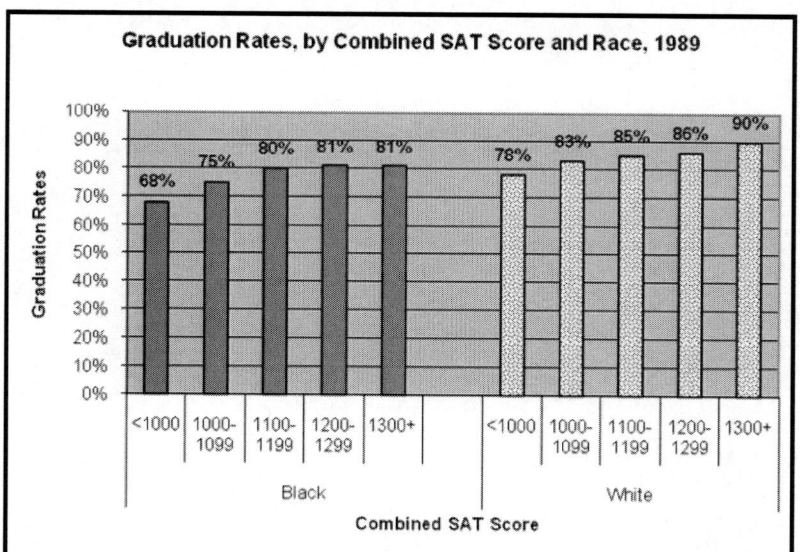

Slide 8

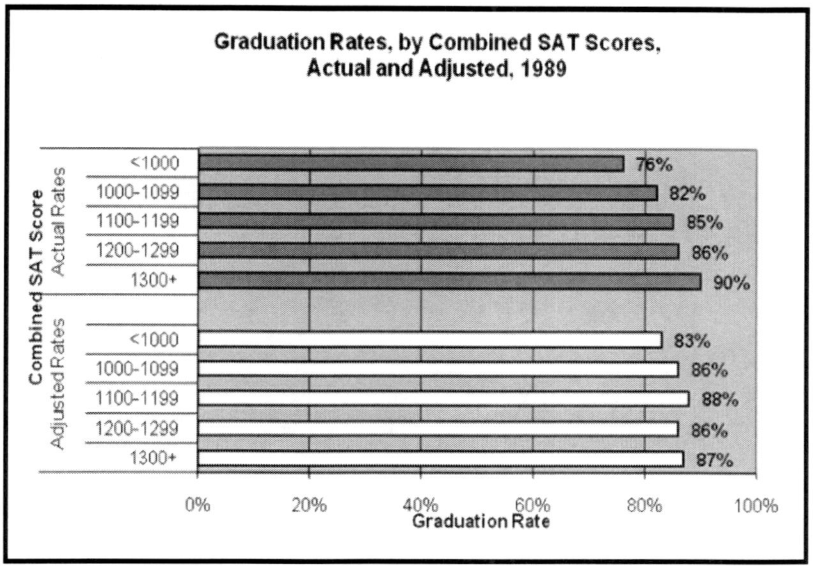


Slide 9

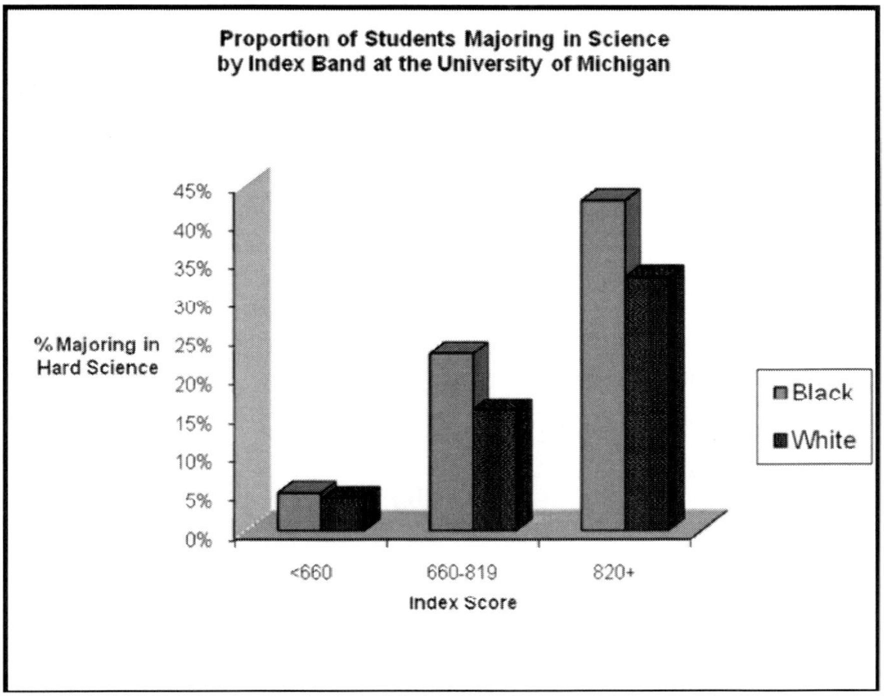

Slide 10

## Michigan 4-Year Graduation Rates

| 1999 Entering Cohort | Large Preference | Moderate Preference | No Preference |
|---|---|---|---|
| Whites | 35% | 52% | 70% |
| Blacks | 21% | 50% | 73% |

Slide 11

## National Longitudinal Survey of Freshmen

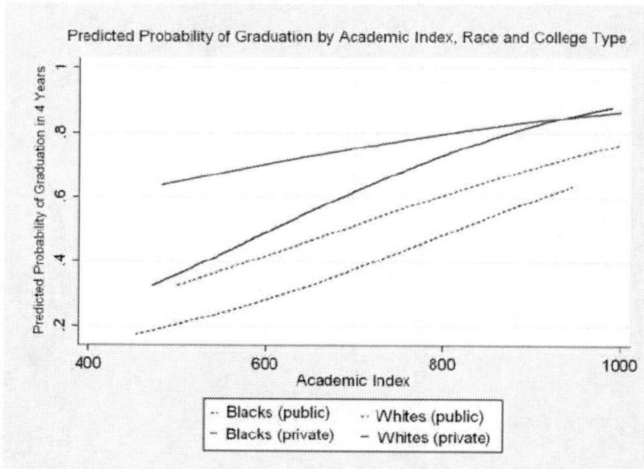


Slide 12

## Hypotheses Three and Four

- From my preliminary examinations of the data, it seems likely that "credential gaps" that exist at the end of high school account for 60-75% of the "science achievement" gap between Blacks and Hispanics, on the one hand, and whites on the other.
- To oversimplify, a central question is whether the rest of the gap is due to poor treatment of minorities in college or to the "mismatch" effects resulting from preferences.
- To test this, we want to compare similar students who experience different levels of mismatch, and we want to compare outcomes that control for race, mismatch, and credentials.

Slide 13

## The University of California data

- In the past six weeks, Project Seaphe has obtained an enormous volume of data from the UC system, covering the period 1992-2006 across eight UC campuses, ranging from the very elite (Berkeley and UCLA) to very good but non-elite schools (Riverside and UC Santa Cruz). It is possible to examine college outcomes – in particular, the outcome of getting a BA in science – by comparing students who are heavily "mismatched" at Berkeley with similar students who attend other UC campuses. These students will be substantially mismatched at UCLA, but only moderately mismatched at other campuses.

Slide 14

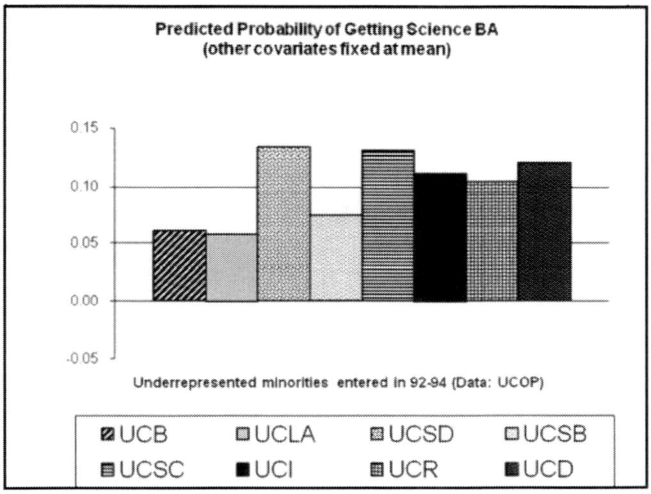


Slide 15

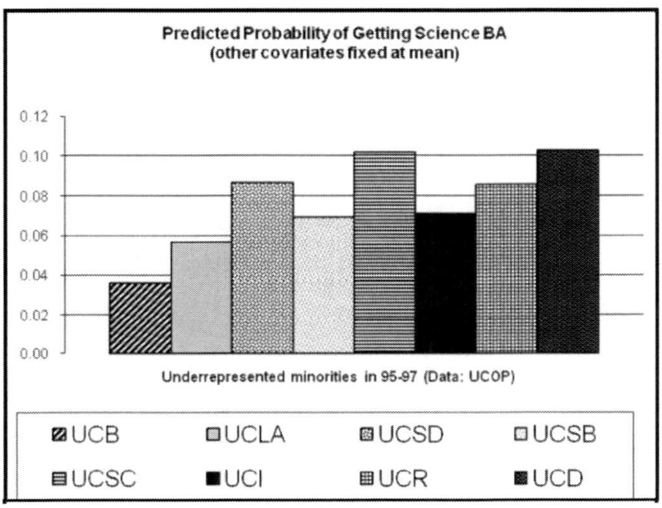

Slide 16

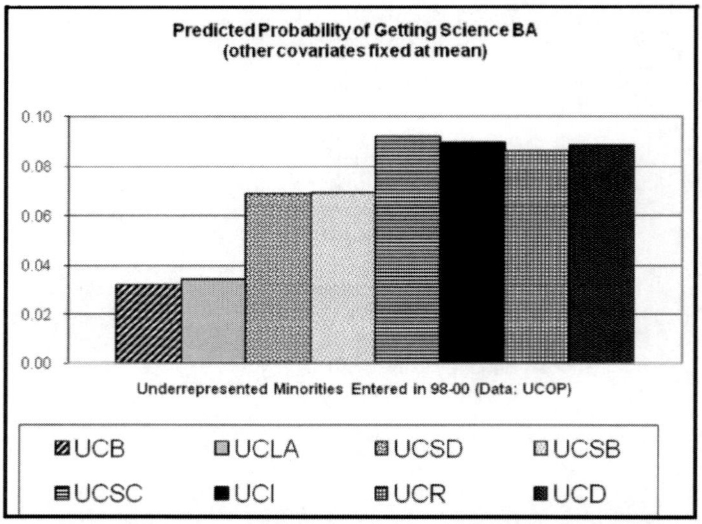


Slide 17

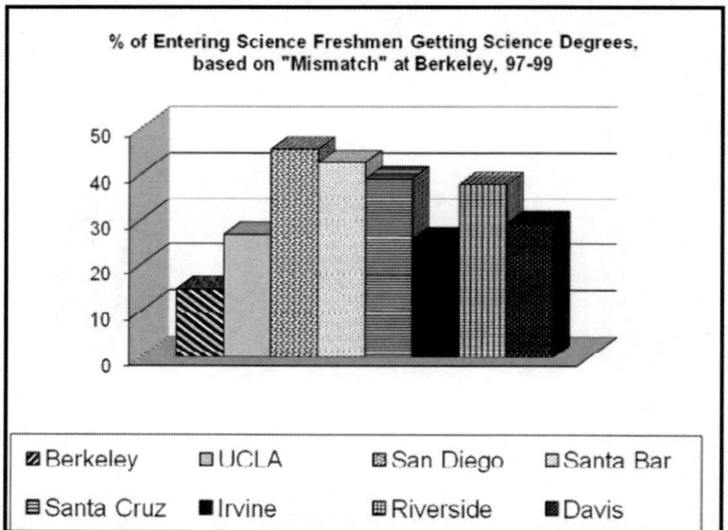


Slide 18

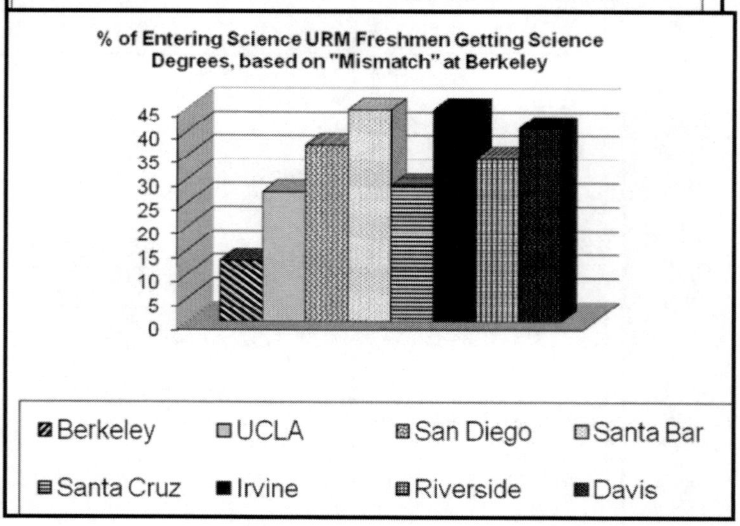

Slide 19

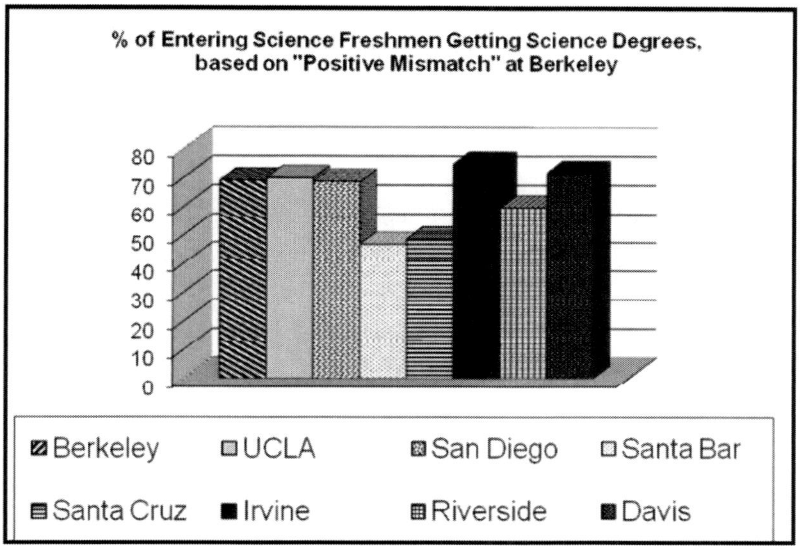


Slide 20

An analysis with CLA data
Examining Predictors of Being a Math Major
As A College Senior

| Variable | Standardized Coefficient | Significance Level |
|---|---|---|
| Intercept | n/a | <.0001 |
| SAT | .19 | <.0001 |
| Mismatch | -.25 | <.0001 |
| Gender | -.03 | .41 |
| Black | .006 | .91 |

## Richard Tapia

Dr. Richard Tapia,[10] Rice University Professor of Engineering and Director of the Rice Center for Excellence and Equity in Education, stated that under his leadership, Rice University has produced what he believes is likely the largest number of underrepresented-minority math, science, and engineering doctoral graduates in the country. Unlike the testimony of Dr. Sander and Dr. Elliott, Dr. Tapia's testimony centered on graduate, rather than undergraduate, education. He noted that Rice University is a highly competitive top research university.

Dr. Tapia recounted his personal history as an underrepresented minority whose parents came to the United States from Mexico, and his early education at a below-average high school in Los Angeles where he was not counseled to attend college despite his mathematical talent. He praised math teachers at the community college he attended for directing him to UCLA, where he earned a doctorate in math. He described his achievements, such as being the first native-born Hispanic elected to the National Academy of Engineering, first minority mathematician promoted to the rank of University Professor, and an appointee to the National Science Board under President Clinton. Dr. Tapia also cited minority STEM graduates who have gone on to distinguished careers.[11]

He stated that the STEM disciplines are a fundamental asset to national economies, and that the United States leads the world in STEM higher education, producing STEM leaders for most of the world's industrial nations. He stated that top research universities choose faculty with doctorates from top research universities, and that professors with doctorates from minority-serving or less prestigious schools are not considered for faculty positions in top schools.

Dr. Tapia did not dispute Dr. Sander's and Dr. Elliott's data supporting the mismatch theory.[12] He did state, however, that in his view, minority students and faculty must have an "equitable presence" at top research universities to serve as role models and mentors. He asserted that including the larger number of Ph.D.s from lesser universities in data showing total numbers of minority STEM graduates would perpetuate the stereotype that minorities receive inferior STEM education. He said the proper form of affirmative action is ensuring that very capable students are not excluded by an overly rigid use of standardized test scores. In his program, he applies a cutoff score, both to exclude those who have little chance of succeeding and to include those who may be on the threshold but will succeed with good mentoring. He stated that applicants with combined math-verbal scores of 1400 or above are well able to succeed, and he does not base admissions judgments on scores that are at or above that level. With regard to cutoff scores, Dr. Tapia agreed with Dr. Sander that there is much more predictive information at the lower ranges of SAT scores than there is at the top, meaning that someone with combined math-verbal scores below 800 will probably not be successful in any STEM discipline. He stated that each year his program rejects several applicants with perfect 1600 SAT scores because test scores are not enough to predict success.

Dr. Tapia claimed success in graduate student retention by providing peer/faculty mentoring and community-building, eliminating deficiencies in preparation, and combating minority students' perceptions of isolation, which he considers the chief problem in many cases.

At the same time, he agreed that admission without retention is of negative value.[13] Moreover, he conceded that most efforts to increase minority presence in STEM graduate programs and faculties have not been effective.

He nevertheless stated that bringing a minority-serving school up to the level of Rice is not as realistic as creating retention programs at Rice that would have the same retention success rate as those at the minority-serving institutions. Dr. Tapia stated that a student with lower STEM grades from a top research school is stronger than a student with top grades from a minority-serving school.

He concluded his remarks by emphasizing that "treating everyone the same is not good enough,"[14] and "if we leave schools alone and don't fix them, then the disaster that is happening will continue."[15]

## Rogers Elliott

Dr. Rogers Elliott,[16] Emeritus Professor of Psychological and Brain Sciences at Dartmouth College, stated that his data supported Dr. Sander's findings concerning mismatch (Dr. Elliott noted that his analysis and data published in 1996 formed an early instance of what Dr. Sander termed "the mismatch hypothesis."[17]) and that race preferences in admissions were harming the aspirations of some black students seeking to be scientists. He asserted that the differences are largest at the most selective universities because of their high admissions standards (two standard deviations above average in developed ability), which some minorities, especially some black students, fail to meet.

Dr. Elliott described his research among 5,300 students in four Ivy League schools between 1988 and 1992.[18] He reported that the black versus white/Asian gap in SAT scores has actually widened since 1988, and is now about 209 points. His data included the students' high school and college transcripts, and SAT scores.

The results (slide 1) showed that an equal portion (about 43 or 44 percent) of incoming students of the three major racial/ethnic groups (with whites and Asians combined as one group) intended to major in science, but that the science persistence rates were considerably lower for Hispanics and blacks: 34 percent for blacks and 56 percent for Hispanics, versus 62 percent for the white/Asian group.[19] These data showed that the average incoming SAT math score for whites and combined white/Asian groups interested in science majors was 715 (the 50th percentile of all incoming students); the average SAT math score for Hispanics was in the 16th percentile, and the average SAT math score for blacks was in the 4th percentile of the white/Asian distribution, a difficult competitive position.[20]

A general college preparedness score called the academic index, which consists of high school grades and rank, and achievement and SAT test scores, showed similar percentiles for the racial groups in Dr. Elliott's data.

Dr. Elliott emphasized that because science knowledge is hierarchical (meaning that a student must understand material in one course before advancing to the next) students who arrived unprepared for the level and pace of work at the highly selective schools in his study were unable to advance successfully through the course sequence compared to the 90 percent of students who were well prepared. His data showed that science grades broken down by SAT math scores reflect this.[21] More specifically, 90 percent of science majors had SAT math scores of 650 or better; of that group, 80 percent were white or Asian students.

Only 25 percent of black students in the four Ivy League schools reached that score. Since the score of 650 appeared to be a break point in the function describing the relation between SAT math scores and the probability of majoring in science (slide 2), blacks were clearly at a disadvantage.

Dr. Elliott's STEM retention data for white/Asian students was 63 percent, 56 percent for Hispanic students (Dr. Elliott viewed this figure as quite good), and 34 percent for black students. He stated that for the year 2008, SAT math score data indicate that only about 3,200[22] black students in the country reached 650, a number he viewed as problematic for STEM recruitment to elite research institutions.

He contrasted the 50 percent chance that someone in the top third of SAT math scores in a very selective school would get one of the science degrees awarded, with the 15 percent chance that a student in the bottom third would get a science degree. At a less selective institution, a student with an SAT score that would be relatively low for a very selective institution would increase his chances of getting a science degree.[23] (slide 3).

Dr. Elliott acknowledged the value of a degree from a top school.[24] He stated that he did not know how to address this dilemma, other than to point out that it is probably better for an aspiring black scientist to go to a school where he or she could get a science degree than to suffer the consequences of relative unpreparedness in a school where he or she was overmatched.

Dr. Elliott's slides are reproduced below.

Slide 1

**Table 2. Revised and Simplified Preadmission and College Performance Measures for Students Initially Interested in Science**

| Measures | White/Asian (N=1782) | | Hispanic (N-95) | | Black (N=157) | |
|---|---|---|---|---|---|---|
| | M | Percentile | M | Percentile | M | Percentile |
| A. Preadmissions | | | | | | |
| % initial interest | 43.3 | | 44.0 | | 44.2 | |
| Number HS Science Courses | 9.95 | 50 | 9.58 | 38 | 9.47 | 35 |
| HS Science Grades | 3.76 | 50 | 3.62 | 33 | 3.44 | 15 |
| HS Other Grades | 3.68 | 50 | 3.51 | 38 | 3.46 | 34 |
| SATM | 715 | 50 | 653 | 16 | 607 | 4 |
| SATV | 626 | 50 | 563 | 20 | 541 | 13 |
| SAT II, ACH | 678 | 50 | 630 | 23 | 573 | 6 |
| Academic Index | 207.2 | 50 | 193.7 | 15 | 182.4 | 4 |
| B. College Performance | | | | | | |
| Science Grades | 2.98 | 50 | 2.46 | 23 | 2.21 | 14 |
| Other Grades | 3.23 | 50 | 2.97 | 32 | 2.85 | 24 |
| % Science Majors | 62.3 | | 55.8 | | 33.8 | |
| % Termination | 4.1 | | 10.5 | | 14.6 | |

Note. "Percentile" refers to the place on the White/Asian distribution of the average member of each group.

Slide 2 (Figures for four Ivy League schools)

**ETHNICITY IN CHOOSING AND LEAVING SCIENCE**

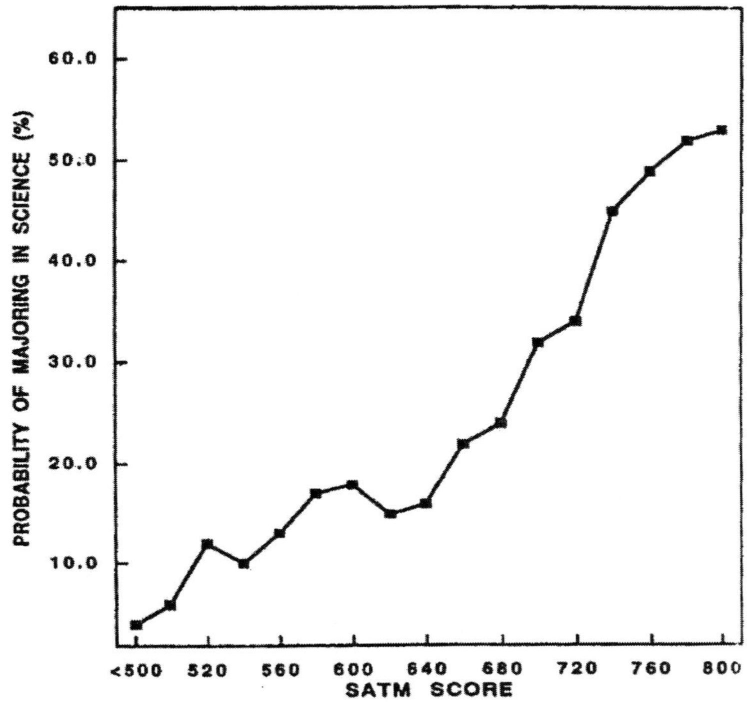


Figure 2. Probability of majoring in science given a particular SATM score.

Slide 3

| | Tercile 1 | | Tercile 2 | | Tercile 3 | |
|---|---|---|---|---|---|---|
| **Institution** | **% Degrees** | **SATM** | **% Degrees** | **SATM** | **% Degrees** | **SATM** |
| Institution A | 53.4 | 753 | 31.2 | 674 | 15.4 | 581 |
| Institution B | 57.3 | 729 | 29.8 | 656 | 12.9 | 546 |
| Institution C | 45.6 | 697 | 34.7 | 631 | 19.7 | 547 |
| Institution D | 53.6 | 697 | 31.4 | 626 | 15.0 | 534 |
| Institution E | 51.0 | 696 | 34.7 | 624 | 14.4 | 534 |
| Institution F | 57.3 | 688 | 24.0 | 601 | 18.8 | 494 |
| Institution G | 62.1 | 678 | 22.6 | 583 | 15.4 | 485 |
| | **Tercile 1** | | **Tercile 2** | | **Tercile 3** | |
| **Institution** | **% Degrees** | **SATM** | **% Degrees** | **SATM** | **% Degrees** | **SATM** |
| Institution H | 49.0 | 663 | 32.4 | 573 | 18.6 | 492 |
| Institution I | 51.8 | 633 | 27.3 | 551 | 20.8 | 479 |
| Institution J | 54.9 | 591 | 33.9 | 514 | 11.2 | 431 |
| Institution K | 55.0 | 569 | 27.1 | 472 | 17.8 | 407 |
| **Medians** | 53.6 | | 31.4 | | 15.4 | |

Note: Percentages indicate the proportion of natural science degrees awarded to students as a function of terciles of the SATM score distribution. SATM numbers are mean scores for each tercile, which vary depending on the selectivity and general level of developed ability that characterizes an institution. SATM is the score on the mathematical reasoning section of the Scholastic Assessment Test.

## Thomas Fortmann

Dr. Thomas Fortmann,[25] who has an undergraduate degree in physics from Stanford and a doctorate in electrical engineering from MIT, provided a teacher's view of the two questions posed by the Commission's briefing topic: 1) why minority students disproportionately leave STEM disciplines, and 2) whether better matching between student and school means that more are likely to succeed.

His answer to the first question was that anyone who arrives in college with inadequate STEM preparation will have difficulty finishing a science major, and that certain minorities frequently have not had adequate K-12 science and math preparation. He described the cumulative nature of math knowledge, which requires that students master information in sequence.

He recounted his efforts to improve STEM teaching by founding an institute to train elementary school teachers in mathematics content. He stated that attempts by programs such as Massachusetts's STEM Pipeline Initiative[26] which attempts to prepare more K-12 students for further STEM study, are well-intentioned, but do not address the root cause of the problem. He offered as an example a simple fractions question given to fifth- and sixth-grade math teachers, only 24 percent of whom could answer it correctly.[27] He stated that Massachusetts will initiate an elementary teacher qualifying test and require college math courses to address this problem by the spring of 2009. Dr. Fortmann said that teacher quality is the most important underlying factor in student math achievement, and that math content knowledge among teachers in urban minority schools is very low.

Dr. Fortmann's view of affirmative action is that, in the absence of sustained remediation as propounded by Dr. Tapia, it does not address the underlying math deficiencies in K-12 education. Dr. Fortmann thus agreed with Dr. Sander and Dr. Elliott that it will consequently hinder some students' STEM progress. Dr. Fortmann argued that it was worth putting the limited resources available into increasing the pool of students able to enter STEM disciplines, rather than just recruiting harder from the same pool. He also recounted a statement by a dean of engineering at a major research state university that 50 percent of the people entering engineering majors eventually switch majors because of inadequate math preparation.[28]

He recommended that the Commission investigate, as a civil rights issue, why so many minority students graduate from high school with so little STEM preparation, i.e., why they need affirmative action at all, and what policymakers can do about it. He believes that the answers are, among other things, school choice; better teacher preparation and more demanding math certification requirements; and differentiated pay scales and incentives. He also mentioned collective bargaining, accountability, standards-based testing, and school leadership as important issues.

## Robin Willner

Ms. Robin Willner,[29] Vice President of Global Community Initiatives at IBM, testified that changes in the global economy make it even more important that the U.S. provide the talent and leadership it needs to remain competitive, and to this end, IBM has begun a STEM program with Latino [Hispanic] students. Ms. Willner stated that other countries have realized that their competitiveness relies on producing and keeping STEM talent. IBM's view is that all sources of talent from all economic and ethnic groups need to be tapped, since the need for STEM graduates will grow by 50 percent, according to a U.S. Department of Labor study. Ms. Willner observed that the U.S. is the only industrialized country that will increase in population in the future, and that much of the growth will come from the Latino community; however, this community currently makes up a very small percentage (1.5 percent) of STEM doctorates. She asserted that the high school dropout rates for Latinos were twice that of blacks and three times that of whites.

Ms. Willner stated that commissioned research papers from Public Agenda, a nonprofit research organization, showed that the Latino educational pipeline is virtually broken in all subject areas, not just STEM, and that the major causes were poverty and lack of English language skills, role models, adequate parental involvement stemming from long work hours, language barriers, lack of formal schooling, and cultural attitudes.

She related some of the actions taken by IBM, such as providing translation programming software and early childhood reading programs, and described a joint meeting on Latino STEM careers that IBM hosted for global companies such as Exxon Mobil, Lockheed Martin, and AMD. The meeting produced four chief recommendations, the second of which bears on the testimony of Dr. Sander and Dr. Elliott: 1) recruit, prepare, and retain qualified STEM teachers, increase the number of second-career STEM teachers, and redesign existing teacher preparation courses and certification; 2) reduce attrition in minority STEM graduates by moving them to programs where they will succeed and also to make sure they can succeed at the highest levels by providing appropriate mentoring, support services, and financial aid; 3) increase the popularity of STEM careers in the Latino community; and 4) increase the Latino high school graduation rate by providing mentors, requiring performance standards of both high school and student, and providing internships.

## DISCUSSION

Vice Chair Thernstrom invited the panelists to respond to each other's testimony before opening the question period to Commissioners.[30]

Professor Sander began the discussion by summarizing the points of agreement among the panelists, such as the academic credentials gap between racial groups by the end of high school as the single biggest cause of the problems under discussion. He observed also that Dr. Tapia and Dr. Fortmann believed that such gaps might be counteracted by good academic support. Dr. Sander stated that most of the panelists did not deny that there was a mismatch problem, but differed on methods of handling it. Dr. Sander also posited that the credentials gap in large part was not caused solely by K-12 education, but by other environmental factors unrelated to schooling, such as low birth weight, parenting practices, socioeconomic differences, and reading habits. He also referred to studies[31] that show that race does not predict credential score gaps if such environmental factors are controlled. Dr. Sander concluded that policymakers should look beyond race to seek out those who need help and end racial preference programs, particularly in the most selective universities whose use of race leads to preferences chiefly for upper-middle and upper-class applicants.

Dr. Fortmann agreed that socioeconomic factors were important, as were parenting practices, and that K–12 education was not the sole cause of the credentials gap. He asserted, however, that there were schools, many of them charter, that improved the educational outcomes for students from disadvantaged backgrounds.[32] Ms. Willner and Dr. Tapia agreed that such schools could make a large difference. Dr. Tapia cited two viable K–12 programs in which he was involved,[33] and stated also that the practice of including all Hispanics in one group was misleading, since educational outcomes differ among subcategories such as Mexican-American, New York Puerto Rican, and Cuban, for example. He recounted coping methods he used as a UCLA student that included taking a reduced course load, and improving his performance in increments rather than all at once.

Commissioner Kirsanow asked the panelists, particularly Ms. Willner, whether minority STEM graduates of historically black colleges and universities (HBCUs) were in fact incompetent by private sector standards, and whether the private sector thought it was important that their new hires come from elite schools such as Harvard, Stanford, and Cornell.

Dr. Tapia asserted that the HBCU STEM graduates were not incompetent, but insufficiently prepared to be competitive with graduates from schools such as Rice. Also important, in Dr. Tapia's view, was recognizing that a degree from a highly selective school is necessary in order to be hired as faculty at a similarly selective school. Part of the problem, he stated, was that some HBCUs are wholly nonselective and are unable to enforce academic standards for fear of failing too many students. Dr. Tapia questioned whether HBCUs were at a point where they should reevaluate their role, since their original purpose resulted from segregation and the nation's colleges are no longer segregated.

Commissioner Kirsanow asked Ms. Willner to comment. She answered that the most important qualification was the applicant's knowledge and skill, not the reputation of the school, but that more selective schools produced more highly skilled and creative graduates. Vice Chair Thernstrom asked whether IBM had its own testing process for applicants. Ms. Willner stated that it did, but that to some extent the company used the reputation of the school as a proxy for such evaluation because of IBM's familiarity with the schools' programs.

Professor Sander addressed Commissioner Kirsanow's question by differentiating between mismatch in undergraduate and graduate STEM programs. Dr. Sander asserted that faculty hires are dependent almost entirely on the graduate advisor and school attended, whereas the science minority STEM dropout problem was an undergraduate problem that could be alleviated by reducing the mismatch between student and school. He referred to social science studies that showed HBCU graduates in both STEM and non-STEM majors who perform well in college have slightly better outcomes over the years than students who perform poorly at elite schools.

Commissioner Gaziano asked Dr. Tapia if he thought it was better to have one successful STEM minority graduate from an elite school and a hundred STEM dropouts, or 100 STEM doctorates at slightly lesser schools. Dr. Tapia answered that 100 successful STEM graduates from lesser schools was a better outcome.

Commissioner Gaziano then asked how large a mismatch could be before it undermined success. Dr. Tapia answered that the question should be why so many top math graduate schools made no effort to retain capable minorities, particularly public universities. Dr. Elliott answered that if the top schools reduced the degree of the mismatch of their minority STEM admits by admitting only the top half of those currently admitted, then schools down the selectivity index would have a greater number of properly matched minority STEM students available to them. Such students would then be at the 30th percentile of competitiveness within their schools, as compared to the fourth percentile. He acknowledged, however, that schools that are less selective also have much less money with which to assist students, and minority students often need such help. Dr. Elliott supported some degree of affirmative action in public universities, but pointed out that at the graduate level, elite schools take very high-level students and push them to meet extremely high standards. The outcome is a high level of performance at such schools, which if lowered, would result in a loss of the school's reputation and elite status, and a loss in performance quality in their programs.

Dr. Tapia agreed with much of what Dr. Elliott had said concerning high levels of performance in elite private schools, but stated that public universities had a moral obligation to do more than just research using the highest performing students available regardless of racial or ethnic imbalances.

Vice Chair Thernstrom asked what effect there would be on K-12 schools with ill-trained teachers if colleges required uniform standards for every applicant; and whether such standards would result in increased pressure on such K-12 schools because they would be unable to ascribe blame to others.

Dr. Sander agreed that the mismatch effect is a problem for those students who receive the largest preferences, while his data concerning those students who receive modest preferences did not show a mismatch effect. He stated that socioeconomic preferences were less likely than racial preferences to produce mismatch effects because those credential gaps were smaller; economically disadvantaged students who persevered all the way to an elite school generally displayed essential elements for success, such as determination and drive. Dr. Sander also asserted that a student coming into a school at the 25th or 30th percentile in terms of credentials instead of at the 5th or 10th percentile could use the resources of the elite school more effectively.

Ms. Willner agreed with Dr. Tapia that the size of the preference (degree of mismatch) is what determines a student's success or failure in a STEM program. She stated that accepting an applicant whose credentials are far below those of the other accepted candidates is different from accepting a student at the top of the SAT scale (1400–1600) with some leeway for modest differences from the median. She agreed with Dr. Tapia that someone at the lower end of the SAT scale with a combined score of 800 would not succeed in a selective STEM program. She also agreed with Dr. Tapia that public universities have an obligation to provide effective support and interventions to students admitted to their programs, and that IBM views as critically important the existence of a diverse group of STEM graduates available to global companies. Dr. Tapia emphasized that it was important for potential minority students to see faculty of their own ethnicity or race because that gave them a sense that they were not alone, and that others of their ethnicity or race had succeeded.

Commissioner Gaziano asked Dr. Fortmann what minimum qualifications math teachers should have; for example, whether second-career STEM professionals need further certification to be able to teach in K–12 schools. Dr. Fortmann responded that Massachusetts allows someone to obtain a provisional license for immediate entry into K-12 teaching, but that requirements differ from district to district. He noted also that such teachers rarely enter elementary or middle schools, so there is still a need to improve the qualifications of existing K-8 teachers. Ms. Willner added that IBM offers its employees a teaching skills program that facilitates the transition to K–12 STEM teaching, because knowing the content area is not enough to become a good teacher. The program, called "Transition to Teaching," helps develop the other essential skills in teaching, and provides a paid leave of absence for student teaching before the employee leaves IBM.

Vice Chair Thernstrom observed that other programs also have been successful in placing non-education majors into classrooms after very short preparation, such as the Teach for America program. Ms. Willner agreed, but added that in her experience, most people who participate in that program claim that the first year of teaching was unnecessarily difficult because of inadequate preparation.

Dr. Fortmann agreed with Ms. Willner's support of more extensive teaching preparation than what is provided in the Teach for America program, but did not believe that an entire year of education courses, or the onerous certification requirements in many states, was necessary.

Commissioner Kirsanow stated that the U.S. Supreme Court's *Grutter* decision[34] set limits on a school's use of racial or ethnic preferences of any size in admissions. He pointed out that however important diverse points of view were in law schools for admissions purposes, and thus arguably within the bounds of the Constitution, the same could hardly be said of physics and math disciplines in which ethnic differences were irrelevant to the quantitative analysis performed.

Dr. Tapia believed that an ethnically broad student and faculty community teaches white students how to deal with people of different ethnicities and races. He also stated that such experience in a broadened academic environment allowed students to build a multi-ethnic academic community that would last throughout their lives. He added that, in his view, court decisions had significantly cut back affirmative action programs, with the result that colleges now recruited from abroad to the detriment of American minorities.

Professor Sander raised two points concerning the practical effects of race preferences. First, he stated that the use of preferences may produce both positive and negative results, meaning that preferences may combat negative stereotypes by the increased presence of minorities, but may also reinforce negative stereotypes if the ability differences are large—on the order of a two standard deviation gap. Second, he stated that large racial preferences further isolate such students socially from normally admitted students, resulting in increased self-segregation along racial lines. Professor Sander supported this statement by citing a recent study posted on Duke University's website by three economics professors at Duke University.[35] He also reported results of his own study of study groups in law school, which found that if there are large racial preferences, whites and Asians assume that the best students to invite into their study groups will be whites and Asians rather than blacks and Hispanics. He asserted that study groups are important in law school because they improve students' grade point averages, and that study groups composed only of minority students do not improve grades.[36]

Vice Chair Thernstrom observed that Harvard undergraduate admissions does not give large racial preferences, but that before the ban on preferences in California, the racial gaps at Berkeley between underrepresented minorities and whites or Asians was enormous. The Vice Chair also observed that the pool of high-SAT-score underrepresented minorities is small, and absorbed by very few schools.

Commissioner Taylor said he found it disturbing that schools were not communicating information about mismatch and its consequences to minority students and their families before they determine which college to attend, and how much success they might achieve relative to other students. He stated that a student needs to know what the range is at any particular school and whether he or she is within that range; if the student is below that range, no additional help from the school or anyone else will be sufficient. He emphasized that the critical public policy perspective should focus on what will help the minority community to do well.

Professor Sander articulated four policies to address Commissioner Taylor's points that he believed all the panelists would endorse. First, he recommended transparency, meaning that African-American or other minority students ought to know before selecting a college what the ultimate outcomes have been for students with their credentials. Commissioner Taylor asked the panelists whether they would agree with Professor Sander's first point, and they replied affirmatively. Ms. Willner added that she thought the colleges also owed students evidence demonstrating that they had given needed help to similar students in the past. Commissioner Taylor endorsed this suggestion and pointed out that this information was no different than providing prospective athletic students a coach's track record of success. He observed that policies of affirmative action had been more concerned with enrolling a target number of black and Hispanic students than helping them make sound decisions about the best colleges for them as individuals.

Professor Sander next suggested that colleges be held accountable to all prospective and current students, meaning a) they disclose their record of retaining science majors to graduation; b) they take responsibility for improving their record; c) they change their admissions practices; or d) all of the above. His third suggestion was that college admissions emphasize socioeconomic preferences over racial preferences. Lastly, he recommended that colleges curtail large mismatches in admissions that have clearly shown negative effects on minority students.

Commissioner Kirsanow asked Professor Sander for his view of the extent of current transparency and accountability. Professor Sander stated that on a scale from zero to 100, it was a 3. Dr. Tapia agreed that getting such information was extremely difficult, and said the recommendation for greatly increased transparency and accountability was excellent. He also gave examples of successful programs, such as Rice's, that were able to recruit able students by word-of-mouth because of a reputation for producing minority successes. He gave as a cautionary example a student who told Dr. Tapia that he had been accepted to Princeton with a 940 combined SAT score, and had been told (and believed) that the average SAT score at Princeton was 950. Dr. Tapia said he urged him to check further.[37]

Commissioner Taylor stated that he strongly opposed any public policy that masked discussion of such information, and further, that he believed concealing such information was intentional.

Dr. Elliott agreed that the size and significance of mismatch in schools was hard to obtain, and that schools concealed such data in order to meet their quotas or goals for enrolling minority students. Vice Chair Thernstrom asked whether colleges cared more about showing off diversity of their freshman classes than results of such policies in senior year with respect to the numbers of minority students in STEM majors. Ms. Willner agreed that letting a student in was easy, but should be matched by a college's willingness to help that student succeed.

Dr. Tapia stated that he visited Berkeley and UCLA frequently and was saddened to see the effects of the California ban on racial preferences there, since administrators had made no effort to create other programs that would work for minorities, such as the "Ten Percent Plan" in Texas. The Texas law requires the flagship campus to accept the top 10 percent of students in each Texas high school regardless of SAT scores, which Dr. Tapia believed has been successful because the university supports the students in special programs once they arrive on campus.

Vice Chair Thernstrom asked Dr. Sander why he thought socioeconomic preferences were less likely to result in mismatch than racial preferences. Professor Sander responded that by broadening the definition of diversity to include socioeconomic preferences, colleges would increase the size of the applicant pool and decrease the size of the mismatch they would have to accept. Second, his data indicated to him that credentials more likely understate potential in lower socioeconomic status (SES) applicants, and they perform better than expected. Third, the resulting smaller gap would not, in his view, create social isolation problems of the kind discussed earlier in the hearing. His experience with expanding admissions to include low SES at UCLA after Proposition 209 passed in California showed that this approach worked. He added that this program was not continued because it didn't produce enough African-American students to satisfy administrators, and that in its place the law school started a "Critical Race" program[38] for minority students admitted with large credential gaps. Professor Sander stated that the Critical Race program was a subterfuge that produced even fewer black law students; during the course of the program, 30 white applicants were denied entry with Law School Admission Test (LSAT) average scores of 163 and eight black applicants were admitted with an LSAT median of 154.[39]

Dr. Tapia interjected a comment that UCLA law school had produced Johnny Cochran; Professor Sander responded that that was an example of entirely race-neutral policies. He added that UCLA's preference programs produced great successes for some individuals, but large preferences had counterproductive effects.

Vice Chair Thernstrom observed that the Bok and Bowen study was of undergraduate admissions only, and that about half the underrepresented minorities had not needed preferences. Professor Sander agreed, noting that Barack Obama had not needed preferences. Dr. Tapia added that the Bowen and Bok study was limited to African-Americans and did not include Hispanics.

Commissioner Kirsanow asked whether schools are discouraging the participation of Asians, who are overrepresented in science programs. Professor Sander said probably not, although he hoped to study that issue. Dr. Tapia stated that to the contrary, U.S. Asians are becoming underrepresented because of assimilation, and schools are bringing in more foreign-born Asians. Professor Sander disagreed in part, referring to his U.S. Census data that showed Asians still hold a large proportion of science doctorates.

Vice Chair Thernstrom asked whether the decreasing portion of Asians in graduate school just showed that they were understandably turning to well-paying business careers, as were other minorities. Dr. Tapia suggested that they were turning to other careers simply because they believed, incorrectly, that a few low grades disqualified them from graduate school.

Commissioner Yaki observed that it is misleading to consider all Asians as one group, since there are great disparities within that group based on assimilation over generations into American society and outcome differences between Japanese or Chinese groups and other Asians. Both Vice Chair Thernstrom and Dr. Tapia agreed. Dr. Tapia stated that a group label is also misleading for those considered "Hispanic" since the disparities in outcomes within such a group are extreme. He stated also that there have been changes in cultural adherence to educational expectations over time within some ethnic groups, resulting in some preference for taking well-paying jobs versus aspiring to attend graduate school, or, in other cases, reduced educational aspirations.

Commissioner Yaki also claimed that since the passage of Proposition 209, the University of California system is accepting foreign Asian students who pay full tuition as a means of avoiding payment of subsidies that accompany in-state students. Professor Sander agreed that the University of California is attempting to maximize its tuition revenues but asserted also that the poorest students in the Los Angeles area are in fact Cambodian, not black or Hispanic.

Vice Chair Thernstrom stated that it is not clearly in the public interest that students enter graduate programs instead of the business world. Dr. Tapia agreed, but asserted that it is bad for University of California and Ivy League math departments to have no African-Americans.

Commissioner Kirsanow asked Professor Sander to what degree the mismatch in STEM disciplines differed from the mismatch in other disciplines, such as law. Professor Sander answered that the STEM mismatch gap at the undergraduate level is uniquely problematic as a result of the linear sequence required for science and math courses, and the impossibility of watering down the required curriculum to push students through to graduation. He observed that law is similar in that it requires bar passage, whereas the academic discipline of English, in his opinion, has no good outcome measures.

Dr. Tapia added that as a result of his frequent work with lawyers as an expert witness in federal court trials, he viewed law and science differently. He claimed that creativity was not a necessary part of success as a lawyer, whereas it is for STEM graduates.

Commissioner Yaki observed that Professor Sander had narrowed his definition of mismatch and that the gap may be due to factors not discussed in this briefing. He also stated that affirmative action in higher education, characterized by Professor Sander as dying, was instead being killed by initiatives across the country. Commissioner Yaki also stated that although he was not criticizing the panelists who were in attendance, he did not believe that they represented a balanced panel on the subject. He further indicated that he was glad that panelist Hazel O'Leary (President of Fisk University and former Secretary of Energy), who had accepted an invitation to participate in the briefing (possibly to speak about the role of HBCUs in STEM) and then cancelled, was not attending this briefing. He stated that the easier accessibility of HBCUs was used as an excuse for doing away with affirmative action, and noted that Dr. Tapia had testified that the HBCUs were insufficient with regard to STEM graduates. Commissioner Yaki stated that affirmative action is about measuring potential, and that closing off access to selective research universities because of statistics would not be acceptable. The briefing then ended.

## STATEMENTS

(Note: Statements are unedited by the Commission and are the sole work of the author. Professor Sander submitted slides in lieu of a written statement that are appended to the summary of his testimony, supra.)

## Richard Tapia

### *The Flaws in the Mismatch Theory*

Picture the nation's research universities in the mid-1960s. Legal and discriminatory obstructions had allowed only a very few underrepresented minorities to gain admission to the nation's top science and engineering institutions. For instance, until 1964, when its board petitioned the courts for a change of its charter, Rice University had been constrained to educate only the "white inhabitants of Houston and Texas." Raymond Johnson then became the first African-American student admitted to Rice and went on to earn his Ph.D. in mathematics in 1969.

His Rice Ph.D. launched a very successful career, including the chairmanship of the Department of Mathematics at the University of Maryland. Stories of early minority scientists and mathematicians produced in this country are amazing, but reflect a truly sad component in our educational history. The few that were produced in those early times invariably were so brilliant with so much potential in their fields that someone in their department would champion their admission, and of course these students were quite successful. But what about all of the other minorities who could have made it, yet had no champions? Certainly this has been a great loss to the country's productivity and leadership.

In the mid-1960s, affirmative action was born. America's universities began to use affirmative action policies to increase the participation of minority groups in higher education. Designed to level the playing field, university admissions policies attempted to normalize for differences in the quality of academic preparation. Admissions criteria were developed to identify students with the capacity to succeed but who were not identified through traditional admissions criteria. Yet these policies have been controversial throughout their history and have faced repeated legal challenges—with much of the controversy centering on whether affirmative action is reverse discrimination and unfair to white students rejected in favor of minority applicants who are perceived as "less qualified."

Recent controversy about the fairness of affirmative action now raises a very different question. Some are now claiming that affirmative action policies can be unfair to the minority students that they are intended to help. The current "Mismatch Theory," promoted by Panelists Rogers Elliott[40] and Richard Sander[41], suggests that minority students are more likely to leave science and engineering when affirmative action has placed them into colleges for which they are not prepared. They contrast this failure with the success that underrepresented minority students experience at less rigorous schools, especially at Minority Serving Institutions (MSIs), and suggest that minority students would be better served by attending less competitive schools where they can be more successful.

It is clear that what many, including several of my colleagues who are underrepresented minorities, want is a strong refutation of the Mismatch Theory as a whole--that it is totally wrong with no foundation or no basis. I claim that the Mismatch Theory is terribly flawed, that it could set underrepresented minorities back 40 years in science participation and achievements, but its flaw is not in its data but in the conclusions drawn by Professors Elliott and Sander. The mismatch theorists have focused light on a huge problem that I have been fighting at Rice for 20 years. The data they present are entirely credible to me because they reflect what I have seen at Rice. So I do not dispute the data. It is the recommendations they make based upon the data that are terribly flawed.

I should explain why I believe that I was asked to serve on this panel. I have been a mathematician at Rice University since 1970. I received a B.A. in mathematics from UCLA in 1961, and a Ph.D. in mathematics from UCLA in 1967. I have received numerous awards for my accomplishments as a mathematician: I was elected to the National Academy of Engineering, appointed to the National Science Board by President Clinton, and at Rice University promoted to the position of University Professor, of which there have been only six named in the history of the university. Upon first glance this appears to look like any traditional academic path to success. We make all kinds of assumptions about the background of such individuals, including the certainty and predictability of the path to success. In my case, most of those assumptions would be wrong.

I was born in Los Angeles to parents who immigrated from Mexico. I attended a below-average high school in the Los Angeles Unified School District. I was not directed to college by my teachers or counselors although I had demonstrated strong mathematical talent. I started to work at a muffler factory where a co-worker recognized my talent and daily insisted that I go to college. I often think of how different my life would have been if this hadn't happened. I was very fortunate. I began at community college, where I was strongly directed to UCLA, and away from less selective four-year colleges, by two of my community college math professors. Little did I know that this advice would be critical to my career. I went on to get a Ph.D. because I saw other students with less mathematical talent than I had who were going on and felt that if they could, I could too.

After receiving my Ph.D. from UCLA, I was directed and guided by David Sanchez, the only underrepresented minority faculty member in the UCLA Mathematics Department, to faculty positions at Wisconsin, Stanford, and Rice. This intervention and guidance was probably the most important in my entire professional life. At many junctures, my life could have taken a very different path rather than to University Professor. Like so many I could have easily fallen through the many dangerously wide cracks. I owe my success to my education at a research university. While at Rice, I have served as dissertation director or co-director for many successful minority doctoral recipients in science, technology, engineering, and mathematics (STEM). Some of them, perhaps most of them, would fit the pattern of the Mismatch Theory, entering Rice less prepared than the majority of their fellow Rice students. Also, I have taught many underrepresented minority Rice undergraduates, and again, some, perhaps most, fit the pattern of the Mismatch Theory. Over the years, I have been frustrated at the number of Rice's underrepresented minority students who migrate from science and engineering to humanities or social sciences, where they have experienced more success.

Selective research universities recruit some of the nation's most capable minority students, who enter intending to pursue a career in science or engineering, and then lose disproportionate numbers of them to other disciplines. I agree with Sander and Elliott that admission of minority students without retention is of negative value. This is what I have, for years, called the "loss of the precious few." Where I strongly part with Sander and Elliott is in what we should do about it. Sander and Elliott say that we should steer students to less challenging schools where they are more successful. According to them, this will be better for the students and better for the nation because it will increase the numbers of underrepresented minority students receiving degrees in science and engineering. I say we should insist that elite research universities put into place programs that have proven successful at supporting students so that they *are* successful. Simply stated, in a "sink or swim," non-mentoring, non-supportive environment, which is what we see at many of our elite research schools, those with poorer preparation will rarely succeed, minority *or* majority. Why are we not demanding from public and private universities that receive federal funds that which is critical for the health of the nation--quality education of all our citizens? Why are we letting them off the hook as they conveniently build an ever increasing base of foreign STEM graduate students and faculty[42]?

### *Creating a Permanent Underclass*

What is wrong with Sander's and Elliott's resolution of this problem? Why do I find it a huge mistake? In my opinion, nothing can do more to establish minorities as a permanent underclass in science and engineering than this.

If the goal is *just* to produce larger numbers of underrepresented minority scientists, then the Mismatch Theory is a great idea, but numbers of degrees alone are not a good measure of success. Underrepresented minorities must be competitive with the overall population; how else can we break the stereotype? The distribution cannot be skewed toward weaker schools. Steering minorities to lesser schools reminds us of the separate but equal mantra. It turns out that separate but equal is always separate, but never equal. But this is worse. This assumes from the start separate and *weaker.* This would take us back to the pre- to mid-60s, where only the very rare minority student who has been prepared well and tests well under traditional admissions criteria would be admitted to the nation's research institutions.

Race and ethnicity should not dictate educational destiny. Steering capable students to lesser schools puts a cap on their potential achievements. Top research universities choose faculty from Ph.D.s produced at top research universities. If we underrepresented minorities are ever to be an equitable presence as faculty at our top-level schools, then our students must be schooled at those same institutions. Leadership in science and engineering comes from top research institutions. MSIs do some things very well. Their students speak warmly of how confident and supported they felt in their experiences there. Research universities should learn from them how to nurture that kind of confidence, but Ph.D.s produced at MSIs will not become faculty at top research universities. We need minorities who will become national STEM leaders, and these have to be produced by institutions that are recognized as giving credibility to the scientific accomplishments of the individual.

Below is a list, which is not exhaustive, that I quickly generated with a few examples of people that I know who received their Ph.D. in science or engineering from an elite research school and who have gone on to assume national scientific leadership.

### *National Leaders in Science and Mathematics who are Underrepresented Minorities*

| **Name** | **Ph.D.** | **Current Leadership Position** |
|---|---|---|
| African-Americans | | |
| Shirley Ann Jackson | MIT | President, RPI |
| Freeman Hrabowski | Illinois | President, University of Maryland Baltimore Co. |
| Shirley Malcom | Washington | Head of Education Programs, AAAS |
| William Massey | Stanford | Professor, Princeton |
| Arlie Petters | MIT/Princeton | Professor, Duke |
| Sylvester Gates | MIT | Professor, Maryland |
| | | Director, Center for String and Particle Theory |

| Name | Ph.D. | Current Leadership Position |
|---|---|---|
| Mexican-Americans | | |
| Hector Ruiz | Rice | Executive Chairman of AMD |
| Rodrigo Banuelos | UCLA | Head, Mathematics, Purdue |
| Francisco Cigarroa | UT/Harvard | President, UT Health Science Center, S. Antonio |
| Richard Tapia | UCLA | University Professor, Rice University |
| Carlos Castillo-Chavez | Wisconsin | Regent Professor, Arizona State |

Where is the list of individuals with Ph.D.s from lesser universities who have outstanding scientific accomplishments or outstanding scientific leadership accomplishments?

***The Systems Are Broken***

Consider three systems that prepare minority students: A) K-12 schools, B) MSIs, and C) research institutions. For very different reasons, none of these adequately promotes equitable representation in science and engineering. But consider, which problem is easier to solve?

A. Transform urban K-12 schools that educate the vast majority of underrepresented students so that they prepare students equally to the best K-12 schools,
B. Bring MSIs up to the academic excellence of research institutions so that capable minority graduates will be competitive with students from elite schools in the industrial job market, professional leadership positions and graduate and professional school, or
C. Design and implement programs at the most selective research universities so that capable minority students have the same retention rates and confidence levels in science and engineering as those at minority serving institutions.

Clearly C has the most viable solution. A by-product of this solution is the added bonus of enhanced training and opportunities and a greater likelihood of ascending to leadership positions. It has taken more than a century to build the sophisticated machinery of research universities.

***So What Do We Do Now?***

Solving the three problems described in A, B, and C above would require a giant overhaul of the entire systems. While we should keep such an objective in view, we cannot wait for this change. There are things that we can do short term that I believe will have a significant impact on improving the representation of underrepresented minorities in STEM careers and leadership positions across the full spectrum of opportunities. My recommendations are as follows.

***Recommendation for Dealing with Problem A***

Talented underrepresented minorities should be identified early in their education (elementary and middle school) and motivated and directed to attend the best magnet secondary schools in the city. This activity would involve working with the parents and the school districts to facilitate and implement these plans. This recommendation has been influenced by the following experience. At the present time at Rice University I am working with three outstanding minority Ph.D. candidates in mathematics and in computer science. All three have distinguished themselves in their research and in their academic accomplishments, including being awarded prestigious National Science Foundation graduate fellowships. Two are African-American and one is Mexican-American. All three were born and raised in the minority areas of large U.S. cities. However, each was directed to a STEM magnet school in their city, performed well, and was encouraged to apply to selective research universities for undergraduate training. All three attended Rice as undergraduates, where I met them and encouraged them to attend graduate school (not necessarily Rice). They received excellent high school and undergraduate preparation and are now outstanding graduate students. I expect to hear in the somewhat near future that they are excellent STEM faculty at research universities in the country.

***Recommendation for Dealing with Problem B***

Many MSIs are open admissions (all who apply are admitted) and also try to be all things to all people. I have a colleague who teaches chemistry at a local Historically Black College and University. He says that in his introductory classes he has some outstanding students and some students who are extremely poorly prepared, and that there is no way that he can do justice to either group of students when they are all in the same class. His level of frustration is extremely high. I recommend that MSIs adopt a magnet secondary school format. They should develop excellent undergraduate courses in selected disciplines and only allow selected, well-prepared students to take these classes. In this way the best students will be well prepared for graduate work in the appropriate discipline at a research university. The details involved in implementing this suggestion would require more thought, but I believe that the direction is correct.

***Recommendation for Dealing with Problem C***

The challenge is to admit underrepresented minority students in larger numbers in science and engineering at the nation's research institutions and then support them to be successful. The research schools must be held accountable for both admission and retention of minority students in their chosen disciplines through the completion of their degrees.

To address admissions, we must evaluate our admissions criteria. I refer to this as second stage affirmative action. Is it excluding individuals with talent to succeed? At Rice, in both graduate and undergraduate admissions, we have successfully turned to what I call the Threshold Approach. We pick a threshold level at which students will be successful that has been determined from years of experience of working with all students. Actually the threshold level is a fuzzy interval of scores. Those students with scores significantly above the threshold are deemed equivalent as far as the test score goes, and the score is dismissed and admission decisions are guided by other factors. Students with scores significantly below the threshold value are not accepted, and those students with scores near the threshold value are looked at with extra care. My experience has been that there is no predictive value at the high end of the test score. For example, there is essentially no value in favoring a student with a combined SAT score of 1500 over one with a combined score of, say, 1300.

The same can be said for a graduate student whose GRE score is in the 95th percentile versus one whose GRE score is only in the 85th percentile. However, I have never seen an undergraduate student at Rice succeed in math, science, or engineering with a combined SAT score below 900.That is, there is much more predictive information at the low end of the scale than there is at the high end of the scale. Indeed, each year Rice rejects a good number (say five or so) of undergraduate applicants who have earned perfect 1600 SAT scores. Of course, the Rice admissions officers feel that some of the high-scoring SAT students were lacking in other significant evaluation components. The misuse of standardized test scores, guided by the belief that there is predictive power at the upper level of the scale, is one of the worst enemies of underrepresented minorities. I have seen underrepresented minority students graduate from Rice with honors, and yet they entered with modest SAT scores, albeit, the highest scores in their minority school. At Rice this is particularly true of Hispanic women.

Following these guidelines, we have produced a very large number, probably the largest number in the country, of underrepresented minority doctoral recipients in mathematics, science, and engineering. In the last 10 years approximately 1,000 STEM Ph.D.s have been produced in the country, and Rice has produced more than 60 of these doctoral students. One year the National Science Foundation informed Rice that in that year we had produced approximately half of the nation's doctoral recipients in the mathematical sciences. The Mathematics Departments at the University of Iowa, through the leadership of David Mandersheid, Cornell University, and Arizona State also have good Ph.D. productivity rates for underrepresented minorities in mathematics; in the latter two situations the champion has been Carlo Castillo-Chavez. Again, success comes from strong commitment, aggressive support, and a champion. These successes demonstrate that it is possible to produce minority Ph.D.s at a high rate at research universities. We now offer a success at the undergraduate level, as well. Due to the Texas Top 10% Rule[43], the Mathematics Department at the University of Texas Austin has the highest percentage of underrepresented minority undergraduate mathematics majors (nearly 30%) of any research university in the country. With innovative support programs they retain minority students through graduation at a rate above the majority student rate.

My summarizing point here is that underrepresented minorities need not be sent to MSIs to succeed. With support and caring we can succeed at the best schools in the country. Indeed, many of us have. And more of us will.

***The Consequence of the Mismatch Theory***

I suspect that many faculty and administrators from research universities would breathe a big sigh of relief when they read about the Mismatch Theory. It certainly lets them off the hook, doesn't it? What it does is reduce expectations and set the country's research institutions back to pre-1964. It ignores all that we have learned about educating minorities and guarantees the formation of a permanent science underclass in America. A two-tiered America is certainly not healthy for the country.

## Rogers Elliott

(Note: Professor Elliott submitted the following article in lieu of a written statement, which may be found at http://www.seaphe.org/pdf/elliott-ethnicity.pdf (accessed July 27, 2010).

***The Role of Ethnicity in Choosing and Leaving Science in Highly Selective Institutions***

From Research in Higher Education, Vol. 37, No. 6, 1996

Authors: Rogers Elliott, A. Christopher Strenta, Russell Adair, Michael Matier, and Jannah Scott

## Thomas E. Fortmann

*Mathematics is ruthlessly cumulative*
*—Steven Pinker*

The Commission poses two questions about science, technology, engineering and mathematics (STEM):

1) Why minority students disproportionately abandon their STEM aspirations,
2) Whether students in institutions matched to their own academic preparation are more likely to remain and succeed in STEM, and to what degree affirmative action policies affect this.

I will attempt to answer the first question and comment briefly on the second, based on my STEM career and recent experience in K-12 education.

My background includes a physics degree from Stanford, a doctorate in electrical engineering from MIT, four years of university teaching, and 24 years as a high-tech engineer and executive. In recent years I have taught mathematics to minority high-school students in Boston, founded a math institute for elementary teachers, and served on the Massachusetts Board of Elementary and Secondary Education.

### *Accumulating Mathematical Proficiency*

Many minority college students avoid or abandon STEM majors for the same reason as nonminorities: insufficient math (and science) preparation in K-12. The disproportionality is surely related to the well-documented lower quality of math/science (and all other) instruction in minority and high-poverty schools. Does affirmative action in college exacerbate that? Perhaps. But the problem begins long before college, and it is most acute in math and math-related fields because math is more cumulative than, say, history or literature.

Steven Pinker made this point rather eloquently when he said:[44]

> "Calculus teachers lament that students find the subject difficult not because derivatives and integrals are abstruse concepts—they're just rate and accumulation— but because you can't do calculus unless algebraic operations are second nature, and most students enter the course without having learned the algebra properly and need to concentrate every drop of mental energy on that. Mathematics is ruthlessly cumulative, all the way back to counting to ten."

I would add that the same is true of algebra: students who don't understand fractions and use a calculator to multiply by ten can't make progress with algebra problems because they're bogged down in the arithmetic (imagine factoring a polynomial if you don't know your times tables).

And so it goes, right down to first grade.

### *Preparation for STEM*

We need more students of every classification in STEM, and in Massachusetts our STEM Pipeline Initiative uses outreach, summer programs, and other marketing tactics to entice more of them, especially minorities and females, into these fields. The problem with this approach is that the "pool" of entering college students who are actually prepared for STEM work is quite small. It would be more cost-effective to concentrate limited resources on increasing the number of STEM-capable students rather than on recruiting harder from the same small pool.

The small proportion of students entering college able to handle STEM courses and majors reflects the dreadful state of mathematics learning in K-12, particularly among minorities. Ask high -school teachers why this is so and they often point to middle school, while middle -school teachers lament students' poor preparation in elementary -school arithmetic. So far no one has blamed the obstetricians.

Unfortunately, this cascade of blame is based in reality. The root problem, in my judgment, is an appalling dearth of mathematics content knowledge among elementary and even middle school teachers; until we solve that, improvements and innovations at the high - school and college levels will have little effect.

Liping Ma was the first researcher to focus national attention on this issue.[45] In fact, my experience suggests that the problem is even worse than she describes. In a group of veteran 5th-6th grade teachers, for example, 57% were unable to answer "75 is 30% of what number?" and 76% could not find two numbers between 1 and 2/5 and 1 and 41/100.

Teachers' attitudes about mathematics range from trepidation to full-blown math phobia, a disease that is highly contagious. This is not the fault of elementary teachers but of preparation programs and certification systems that virtually ignore mathematics.[46,47]

And it creates a vicious cycle in which each generation of teachers is recruited from a group that leaves high school with weaker math knowledge than the last. Massachusetts has begun to address this issue with regulation changes that require more college math courses and a math test for elementary teacher certification.[48]

It's well known that teacher quality is the most important factor driving student achievement, and that teacher quality—including math content knowledge—is generally lowest in poor urban districts serving minority populations. Thus the disproportionality that the Commission cites is no surprise.

### *Affirmative Action*

It seems self-evident, given the cumulative nature of mathematics, that affirmative action will hinder some students' STEM aspirations unless it is accompanied by vigorous and sustained efforts to raise their mathematical capacities to the level of their classmates.

In the absence of effective remediation, should those students be placed in colleges where their peers' proficiency in mathematics is similarly lacking? Doing so may result in more STEM majors, but it masks their (and their peers') low achievement and limited career opportunities. The place to address those crippling math deficits is back in K-12, where poor instruction and math-challenged teachers, along with low standards and social promotion, allow too many students to reach high school and graduate without the mathematical skills and understanding they need.

In other words, affirmative action comes into play late in the game and doesn't address the underlying mathematical deficiencies that deflect minority (and other) students from STEM majors and careers.

***Recommendation***

I encourage the Commission to investigate, as a civil rights issue,[49] why so many minority kids arrive in college unprepared for STEM majors—i.e., why they need affirmative action at all, and what we can do about it. The answers—far beyond the scope of today's briefing— will get to the heart of education reform and teaching quality. There are numerous worms in that can, including school choice, teacher preparation and certification, professionalization of teaching, career ladders, differentiated pay scales and incentives, collective bargaining, accountability, standards-based testing, and school leadership.

That is a tall order, but necessary in order to get beyond the symptoms and treat the disease.

## Robin Willner

Let me congratulate the Commission for convening today's discussion and focusing attention on the very critical issue of minority participation in Science, Technology Engineering and Math (or STEM) careers.

I'm honored and pleased that you have asked IBM to participate today. And while I would not hazard to speak for an entire industry or the private sector, I do believe that my comments would gain a good deal of support from my colleagues. There is widespread concern about the future labor needs in the growing areas of the economy -- every major corporate leader recognizes that the U.S. labor force must continue to provide the talent and leadership that we need for a robust economic future and to remain competitive in a global economy. I also want to take this opportunity to tell you about a recent initiative at IBM that focuses on the Hispanic community and share what we've learned about both the challenges and potential solutions.

Before addressing today's topic, let me briefly provide some important context. While it has always been important for business to nurture and prepare a workforce in the U.S. with the necessary skills, there have been profound changes in the global economy that make this more important than ever before. In a global economy, the world is not only smaller; it's full connected, or, if you will, networked.

There will always be some businesses that will not need to worry about what is happening around the world, more and more, successful businesses will take the form of what we at IBM call the Globally Integrated Enterprise. U.S. companies like IBM create great opportunities for American workers and generate important economic return in the U.S. precisely because we are globally integrated and functioning on a global scale.

In a connected world, we have access to huge new markets. We also can organize our business around the globe to optimize operations. If everything is connected, then work flows throughout the network. And the most important consideration is TALENT. Localities, states and nations are striving to become places where knowledge is generated and transformed into new commercial and societal value. They recognize that, of course, we need employees with basic skills; but more importantly, we need employees who can innovate. A knowledge-based society creates jobs, raises living standards and generates growth that competitors can't duplicate rapidly.

The real wealth generators in today's global economy are technical people. A recent report by the U.S. Labor Department suggests that over the next 10 years, the need for technical people in this country is going to grow by 50 percent.

At the same time, there is a serious shortage of professionals and students studying the fields of science, technology, engineering and mathematics. I won't repeat the detailed statistics that you've heard from other speakers and that your own preparatory research uncovered. But let me be very clear on this point. There are plenty of reasons to make sure that every child in this country has the opportunity and access to be prepared to be an engineer, scientist or mathematician or whatever their dreams dictate. But more importantly, this is an issue of our economic survival. If the key to prosperity is having the right talent, then we must take advantage of the gifts and promise of every child. We ignore any community at our peril. If the U.S. is to remain competitive, we need children from every ethnic and economic background prepared for STEM studies and potential careers in the STEM disciplines.

In our efforts to nurture all of this nation's talent, we need to attend to students from every minority group and young women as well. American ingenuity has always been the product of our rich and diverse heritage and the foundation for our economic strength. Each group has distinctive challenges and opportunities, and IBM has a range of programming to serve the full range of students. However, with our Latino population, we have the classic paradox of challenge and opportunity. Or, as our engineers describe it, "the tyranny of large and small numbers." And let me use a few compelling statistics to explain.

Over the next 40 years, the United States will be the only developed country that will grow its population. Much of that growth will come from the Latino community, who are estimated to constitute 25 percent of the total U.S. population – a growth rate of 30 percent. This segment of the population will certainly be relied upon, heavily, to help drive our nation's economic future.

However, current numbers of Latinos in STEM degree attainment and careers show that they are grossly underrepresented. According to the National Action Council for Minorities in Engineering (NACME), Latinos accounted for only 4.2 percent of engineering degrees awarded in 2005, and merely 1.5 were awarded doctorate degrees. Dropout rates among Latino youths are the largest at 24 percent compared to 12 percent for African-Americans and 7 percent for non-Hispanic whites.

The reasons behind these statistics are extremely complex and deeply entrenched.

That is why IBM, along with Exxon Mobil and Lockheed Martin and 150 leaders from education, business and the community, convened at the *America's Competitiveness: Hispanic Participation in Technology Careers Summit* in May. The objective was to address this once-hidden, but now clearly visible crisis and, more important, to take action. We also commissioned a number of research papers from Public Agenda and The Tomas Rivera Policy Institute to enhance our understanding and help shape our agenda for action.

Perhaps the most compelling information came from Public Agenda, after conducting a series of interviews with national leaders from every sector. The title of their report tells the story: "Out Before the Game Begins."

"Nearly all of the interviewees said that when it comes to Hispanic and Latino students, the educational pipeline is all but broken. Respondents across the board believed that the current educational system is not serving the Hispanic population well. This failure extends to all subject areas, not just science and math. Before these specific subjects can be taught well, most said, the nation needs to bring basic education up to par. According to nearly all of those we spoke with, the overall poverty of Hispanic-Americans is perhaps the largest contributing factor to poor quality education; Hispanics tend to live in areas of concentrated poverty with struggling public schools and a less-than-adequate tax base for funding them. A wide swath of the Hispanic population also lacks the necessary English language skills."

As a first step, we need more teachers that are qualified to teach math and science. For example, Education Trust-West found that 44 percent of math courses at high-poverty-level high schools – where there are large concentrations of underrepresented groups - were led by teachers without mathematics certification. Given this, it's no surprise that these schools are not offering the challenging math and science preparation that students need. In 2006, IBM announced Transition to Teaching, our initiative to address the K-12 STEM teacher pipeline issue by leveraging our greatest asset – IBM employees.

But it's not just an educator problem. The absence of role models for Latino students is a major inhibitor, according to our research. Parent involvement is a factor as well. Immigrant parents face several obstacles that include long work hours, language barriers, lack of sufficient formal schooling, and cultural attitudes carried over from home countries that hinder a parent's ability to be an advocate for their child.

IBM has made a commitment to focus a variety of education initiatives on schools and communities serving Latino students -- for example, offering free automatic translation software to any school district to improve the communication between Spanish-speaking parents and English-speaking teachers, and increasing our mentoring programs so students can interact with role models. I'd be happy to discuss any of these in more detail, and I've provided more information with my written testimony. But, these are problems that cannot be solved by one company or one sector alone. Clearly the private sector, no matter how much it contributes, is only one part of the solution. There were four key recommendations from the meeting.

The first recommendation is to recruit, prepare and retain qualified math and science teachers. We need to create and fund new career paths that encourage the best and brightest to leverage industry experience to enhance their classroom skills and vice versa while developing more competitive salaries with a cross-industry career. In the short term, we need more second career teachers from the ranks of our math and science professionals. The private and public sectors must collaborate to develop financial incentives for tuition, in-service professional development and competitive salaries. At the same time, we need to redesign current teacher preparation programs, encouraging – indeed, demanding that – universities, state education departments, school districts and teacher unions work together to prepare and support excellent teachers.

The second recommendation is to find ways to reduce undergraduate attrition rates for Hispanics in STEM majors. We need to focus on those young people who have expressed an interest in STEM careers and made it as far as a community college, college or university program and surround them with the necessary mentors, support services and financial aid to stay the course and succeed. Other supports could include internships that expose them to

STEM careers, job placement services and other connections into the private sector to start their careers.

The third recommendation is to increase the popularity of STEM careers in the Hispanic community. The public and private sector should sponsor a major marketing campaign that educates the Latino community on exciting and lucrative STEM careers, along with information on how to attain them.

The fourth recommendation is to increase the Hispanic high school graduation rate prepared for STEM career training. Every high school must offer the same challenging curriculum we offer at the most successful schools, and all of the students must participate. As part of this, we must ensure that STEM education starts early and continues throughout a student's schooling. One compelling idea is to establish a formal certification for schools that offer an effective STEM program and meet a standard for student achievement across all socioeconomic and ethnic groups. Public recognition and financial incentives would encourage all high schools to strive to secure the certification. Middle and high school students should have mentors from industry who embody the best that STEM has to offer, as well as internship opportunities that encourage students to dream big and work hard.

Finally, we must find ways to eradicate cultural barriers – such as language – that prevent Latino parents from participating in their children's education.

America's goal must continue to be to raise the standard of living for our children. To do so, we must take aggressive action. We must capture more minds, more hearts, more souls, more passion for the STEM disciplines if we are to retain our competitiveness and attain greater heights of leadership. It is an economic imperative as well as a moral imperative. Whether you are in business, education or a community leader, I encourage you to get involved today.

## Speaker Biographies

### Richard Sander

Professor Sander received his bachelor's degree from Harvard and a law degree and doctorate in economics from Northwestern. He has taught at UCLA's School of Law since 1989, where he does empirical research on social policy. Sander has done major studies of inner-city banking, housing segregation, central city poverty, and living wage laws. He is probably best known for his research on legal education; in this area, Sander has studied academic support programs, class-based affirmative action, the third year of law school, the market for lawyers, and, most recently, the systemic effects of racial preferences in legal education and law firms. Sander has worked to improve the enforcement of fair housing laws in southern California, and helped to start a program that substantially increased the participation of low-income Los Angeles workers in the Earned Income Tax Credit program.

## Richard Tapia

Dr. Tapia currently serves as the Maxfield-Oshman Professor in Engineering at Rice University. While at Rice, he has directed or co-directed more underrepresented minority and women doctoral recipients in science and engineering than anyone in the country. He has received numerous national awards, including the National Science Foundation's inaugural Presidential Award for Excellence in Science, Mathematics, and Engineering Mentoring for his work with these students. Dr. Tapia is recognized as a national leader in diversity and for many years has received dozens of requests to give invited addresses, serve on university diversity committees, and provide leadership at a national level. Leading professional organizations have recognized Dr. Tapia’s contributions to diversity by naming two professional conferences in his honor, describing him as a seminal figure who inspired a generation of African-American, Native American and Latino/Latina students to pursue careers in mathematics.

## Rogers Elliott

Dr. Rogers Elliot is Professor Emeritus of Psychological and Brain Sciences at Dartmouth College. He has served as Chairman of the Department of Psychological and Brain Sciences (1983-86), Chairman of the Department of Education (1975-1980), and as Chairman of the Master of Arts in Liberal Studies Program (1990-1992). He retired from the College in 2001, but has continued to teach a course each year, alternating between psychology and law and individual differences in abilities, with an active appointment as Research Professor. He is the author of Litigating Intelligence, which addressed legal challenges to intelligence tests. With colleagues such as A.C. Strenta, he has studied the issue of student interest in and persistence in science, as functions of gender and race/ethnicity. Dr. Elliot went on to finish his Juris Doctor degree in 1986.

## Thomas Fortmann

Tom Fortmann received a B.S. in Physics from Stanford University and a Ph.D. in Electrical Engineering from M.I.T. After four years teaching at Newcastle University in Australia, he spent 24 years as a high-tech engineer and executive at BBN Technologies in Cambridge, MA. Since his retirement from BBN, he has spent over a decade teaching mathematics as a volunteer in two Boston high schools, founding a math content professional development program for elementary teachers, and working on math/science policy issues in the Commonwealth of Massachusetts. He was appointed to the Massachusetts Board of Elementary and Secondary Education in 2006. He was instrumental in modifying state regulations to require a mathematics test for elementary certification, and he wrote the state guidelines specifying college mathematics courses for elementary teacher candidates. He is the author of an undergraduate textbook on linear control systems, a graduate textbook on multitarget/multisensor tracking, and numerous papers and articles.

## Robin Willner

Robin Willner is Vice President of Global Community Initiatives for the IBM Corporation. She joined IBM in March 1994 to design and implement *Reinventing Education*, a philanthropic initiative in K-12 school reform. This $75 million global program now includes 25 grant partnerships with school districts and states throughout the United States plus 12 other countries, each focusing on a collaborative effort to develop new applications of technology to overcome common barriers to school improvement and increase student achievement. In addition, Ms. Willner oversees a range of global philanthropic and volunteer programs, including the Global Citizens Portfolio, Corporate Service Corps, World Community Grid, Transition Programs, online mentoring, literacy and work force development projects, school-to-career programs and other efforts to apply technology to specific societal issues. She also manages IBM's humanitarian response to disasters, including IBM's response to the China earthquakes, 9/11, the December 2004 tsunami in south Asia, Hurricane Katrina and dozens of others. Prior to joining IBM, Ms. Willner served for three and a half years as Executive Director for Strategic Planning for the New York City Public Schools. In that position, she was the chief policy advisor to Chancellor Joseph A. Fernandez and oversaw all evaluation, research, testing and data collection activities in the nation's largest school district. As Deputy Executive Director of INTERFACE, a New York City-based not-for-profit agency, for more than a dozen years, she directed over 100 research reports on public policy in the areas of education, job training and child welfare. Ms. Willner serves on the Boards of Directors of Grantmakers for Education and the Center for Education Policy, and the National Academy of Engineering's K12 Task Force.

# COMMISSIONER STATEMENTS AND REBUTTALS

## Statement of Commissioner Gail Heriot

The assumption behind the fierce competition for admission to elite colleges and universities is clear: The more elite the school one attends, the brighter one's future. That assumption, however, may well be flawed. The research presented at this briefing provides strong reason to believe that attending the most competitive school is not always best--at least for students who aspire to a degree in science or engineering.[50]

Majoring in science or engineering can be difficult.[51] As one Yale University student told the *Wall Street Journal*, the science course he took "scared the hell out of me." "In other classes, if you do the work, you'll get an A," he complained. "In science, it just doesn't work that way."[52]

Well ... yes ... the feeling that one is flailing about in science or engineering courses can be very disconcerting. Many students who start out with such a major switch to something easier. Others drop out or even flunk out. And it should surprise no one that those who fail to attain their goal of a science or engineering degree are disproportionately students whose entering academic credentials put them towards the bottom of their college class.[53] Not all stereotypes about science and engineering students are accurate. But the basic notion that they tend to be highly-credentialed and hardworking is largely on target. They have to be.

What some do find surprising is this: *Part of the effect is relative*.[54] An aspiring science or engineering major who attends a school where his entering academic credentials put him in the middle or the top of his class is more likely to succeed than an otherwise identical student attending a more elite school where *those same credentials* place him towards the bottom of the class. Put differently, an aspiring science or engineering major increases his chance of success not just if his entering credentials are high, but also if those credentials compare favorably with his classmates'.[55]

The reasons for this comparative effect are doubtless complex. But they are based on a common everyday observation: A good student can get in over his head and end up learning little or nothing if he is placed in a classroom with students whose level of academic preparation is much higher than his own, even though he is fully capable of mastering the material when presented at a more moderate pace. Discouraged, he may even give up–even though he would have persevered had he been in a somewhat less competitive environment.[56]

Science and engineering are ruthlessly cumulative. A student who has difficulty with the first chapter in the calculus textbook is apt to have difficulty with the second, third and fourth chapters. Indeed, the subsequent courses in the mathematics curriculum may be a problem. By contrast, an English literature student who simply fails to read the Chaucer assignment is not necessarily at a serious disadvantage when it comes to reading and understanding George Eliot. Since quitting science and engineering is easy–ordinarily all one has to do is switch majors–the attrition rate is quite high. By senior year, there are significantly fewer science and engineering majors than there were freshmen initially interested in those majors.

Some call this comparative effect the "mismatch" effect.[57] And although there is reason to believe that it applies to other kinds of learning, science & engineering examples are perhaps the easiest to imagine: I have every confidence that I can learn basic physics, despite the fact that I have never taken a course in it and my mathematics skills are a little rusty. If I ever lose my job as a law professor, I suspect that I am fully capable of re-tooling as a physics teacher if that is where the available jobs turn out to be. But if I were thrown into the Basic Physics course at Cal Tech, with many of the very best science students in the world, I would be lost and likely learn little if anything. I would be mismatched.[58] On a good day I might make a few lame jokes about my unhappy situation; on a bad day I might even get a little testy about it. But I would be unlikely to come out of that class as competent in the basic principles of physics as I would have in a less high-powered setting.[59]

That doesn't mean, however, that those who aspire to a career in science or engineering must graduate from high school already prepared for the rigorous science curriculum at the world's most competitive science-oriented university. There are many careers in science and engineering. Many have been filled by latecomers to these fields. It simply means that for those who are not already well-prepared when they begin to study science or engineering in earnest, the best strategy may be to avoid going immediately head-to-head with better prepared students.

The interest of the Commission in mismatch centers mainly on its effect on members of underrepresented racial minorities–primarily African-Americans, Hispanics and American Indians. Since admissions standards are frequently relaxed in order to admit a more diverse student body, minority students constitute a disproportionate share of the students with entering academic credentials towards the bottom of any particular class.[60] Obviously, however, there are other categories of students, such as athletes, children of alumni and other special admittees, who should also be mindful of the risk of mismatch that comes with preferred treatment in admissions.

All such students have a dilemma to face. Should they accept the supposed "leg up" they have been offered? Or should they reject it and attend a school at which that leg up would have been unnecessary? The answer is likely to vary from student to student and may be a question of priorities. Which is more important–that student's desire to attend the most elite school or his or her desire to be a physician, engineer, or scientist?

The problem is that few students who receive a preference realize that their entering academic credentials are well below the institutional median. Fewer still realize that relatively low academic credentials are likely to handicap their ability to earn a degree in science or engineering there and that their odds would be better elsewhere. Instead, they are recruited, indeed romanced, by colleges and universities who allow the scramble for a racially diverse campus (or a winning football team or happy alumni) to overcome their commitment to full and fair disclosure.

It is for this reason that the Commission has recommended that colleges and universities inform the students they are attempting to recruit of the mismatch issue and its potential impact. Tuition for the 2010-11 academic year at Duke University, the University of San Diego and Yale University, for example will be $36,065, $36,950 and $38,300 respectively. That, of course, does not include room and board, books or various student fees. Many students are willing to incur such debt because they envision that their future career will be in a well-paying field like medicine, dentistry, or nuclear engineering. When they graduate four years later with a degree in a soft major, they may be saddled with a much larger debt than they would have been willing to undertake had they understood that the odds were stacked against their success in science or engineering. But no one told them.

At minimum, this is an issue that students should be informed of so that they, with assistance from their parents, high school teachers and guidance counselors and other advisors, can decide for themselves how to proceed. But let's look at the evidence step by step.

### *A. Minority Students Are Indeed Underrepresented in Science and Engineering*

There is no segment of the labor force that proportionally reflects the nation's demographic profile. Physicians are disproportionately Jewish. Jockeys are disproportionately Hispanic. The wine industry employs more than its share of Italian Americans. Even within professions, disproportionality is the rule, not the exception. Among lawyers, litigators are often Irish-American. Among physicians, radiologists are disproportionately Subcontinent Indian-American.

Lack of proportionality is not necessarily the result of systematic discrimination. There are many ways in which one's family background, language, and cultural traditions directly or indirectly affect career choices. As a result, it would be hard to find a single profession or occupation that looks, as it is often put, "like America." The world is always more complex than that.

But science and engineering are special. For one thing, they are not single fields. Instead, obtaining an initial degree in a field of science or engineering is the gateway to a large number of respected professions and occupations--aviation inspector, bio-chemist, computer software engineer, dentist, electrical engineer, forester, geophysicist, hematologist, insurance actuary, jet propulsion engineer, lighting engineer, mathematics teacher, nuclear engineer, optometrist, pharmacist, quantitative analyst, radiologist, science teacher, thoracic surgeon, urologist, veterinarian, waste engineer, x-ray engineer, and zoologist. These fields represent a significant portion of the most lucrative and dynamic sectors of the world economy. If African-Americans, Hispanics and American Indians are facing significant impediments in entering these fields, that is a situation that calls for attention.[61]

Using data from the National Survey of College Graduates conducted by the U.S. Census Bureau, UCLA law professor Richard Sander and senior statistician Roger Bolus have calculated the following racial gap in science among college graduates, including immigrants educated or partly educated abroad, age 35 and under:

**Table I. How Significant is the Racial Gap in Science?[62]**

| Frequency Relative to Population | White | Black | Hispanic | Asian |
|---|---|---|---|---|
| Gen. Pop. | 100 | 100 | 100 | 100 |
| Bachelor's Degree Science | 100 | 36 | 41 | 454 |

As the chart indicates, blacks and Hispanics are only 36 percent and 41 percent respectively as likely as whites to have a bachelor's degree in science or engineering. An Asian, by contrast, is more than four and a half times more likely than a white to hold such a degree. Blacks are only 15 percent as likely as whites and Hispanics are only 26 percent as likely as whites to have a Ph.D. in science. Asians, on the other hand, are more than seven times as likely as whites. The underrepresentation of blacks and Hispanics in science and engineering is real (although these figures are in part a reflection of the immigration of highly-qualified individuals from abroad).[63]

Of course, concern over underrepresentation in science and engineering is not new. On November 13, 1992, the popular magazine *Science* issued a special news report entitled "Minorities in Science." In it, the editors lamented:

> For 20 years, science has been wrestling with "the pipeline problem": how to keep minorities from turning off the obstacle-strewn path to careers in science, mathematics, and engineering. Thousands of programs have been started since the late 1960s to bring diversity to the scientific work force. But their results have been dismal....[64]

One thing, however, is clear. The problem has not been an unwillingness to spend money. By 1992, the National Science Foundation had already spent over $1.5 billion on programs designed to increase the number of minorities in science or engineering. Officials at the National Institutes of Health estimated that they had pumped an additional $675 million into the system. Uncounted state, local, foundation and industry programs contributed millions more.[65]

But the consensus of opinion has been that much of the money had been spent unwisely. In their eagerness to qualify for the vast grants available to educate future minority scientists and engineers, many colleges and universities admitted minority students with little background in science or mathematics. In the early days of affirmative action, "colleges took any person of color who wanted to become an engineer, regardless of their background," said Mary Perry Smith, a former Oakland schoolteacher and founder of California's Mathematics, Engineering, Science Achievement (MESA) program, which promotes minority student participation in those fields. "They tried to turn students who barely knew algebra into engineers and it was a total failure."[66]

"The country cannot repeat the experiment of the last 20 years," said Luther Williams in 1992, then the assistant director of education and human resources at the National Science Foundation. Williams, who later went on to become provost of Tuskegee University, a historically black university with a reputation for emphasizing a science and engineering curriculum, was blunt: Those vast expenditures were "an incredible waste of financial and human resources."[67]

Was Williams being too harsh? Perhaps. Progress has been made and it will continue–even though it is not as much progress as we would like. But if the problem is going to be solved, it will not be solved by more of the same thinking that has characterized the efforts of the last forty years. A re-examination of the assumptions behind those efforts is in order–even if it will step on a few well-entrenched toes.

### *B. There Is No Problem with Lack ofInterest in Science and Engineering Among Minority Students. It Is Disproportionate Attrition that Is the Cause for Concern*

The problem with minority underrepresentation in science and engineering is not the result of lack of interest among college-bound African-Americans, Hispanics and American Indians. Study after study has found just the opposite.[68] Indeed, if anything, such students are slightly more interested in pursuing science and engineering degrees than white students. For example, Professors Alexander W. Astin & Helen S. Astin of UCLA's Higher Education Research Institute examined a sample of 27,065 students enrolling as freshmen at 388 four-year colleges in 1985. They found that the rate of initial interest in majoring in a biological science, a physical science or engineering was, in descending order, 52.6 percent for Asians, 35.7 percent for Chicanos, 34.5 percent for American Indians, 34.2 percent for African-Americans and 27.3 percent for whites.[69]

These findings were consistent with later efforts to study the issue. When Dartmouth College psychology professor Rogers Elliott and his co-investigators looked at a sample of 4,687 students enrolling at four elite colleges and universities in 1988, they found that 55 percent of the Asians, 44.2 percent of the African-Americans, 44 percent of the Hispanics, and 41.4 percent of the whites were initially interested in majoring in science.[70] Similarly, Richard Sander & Roger Bolus, in analyzing all students enrolling in the University of California between 2004 and 2006, found that 57.1 percent of Asians, 40.5 percent of AfricanAmericans/Hispanics and 34.7 percent of whites declared an intention to major in science or engineering.[71]

To my knowledge, no one who has examined the evidence disputes these figures. They are as solid as any figures in social science. If there is a problem with lack of interest in science and engineering, it is with college-bound whites, not African-Americans, Hispanics and American Indians.

To be sure, that doesn't mean that there is no point in encouraging even more underrepresented minorities to aspire to careers in science and engineering. Programs that are proven to encourage such interest may well be money well spent. But if one wants to understand the root of the problem, one must look elsewhere.

And some researchers have. Their work has shown that the problem for minority college students comes a little further down the pipeline. While African-Americans, Hispanics and probably American Indians have high rates of initial interest relative to whites, they are less likely to follow through with that interest. Somewhere in college, the intention to graduate with a degree in science or engineering dies. Alexander & Helen Astin report, for example, that while 68 percent of Asians and 61 percent of whites in their sample followed through on their intention to major in biological science, physical science or engineering four years later, only 47 percent of African-Americans and 37 percent of Hispanics did the same. The rest had apparently changed majors, dropped out, or flunked out.

Consequently, while one might expect, given their level of interest, that African-American college students would be somewhat over-represented among science and engineering college graduates, they turn out to be underrepresented instead. Hispanics are a special case. With them, English mastery is sometimes a problem. One would therefore expect very high perseverance in science and engineering, since transfer to a discipline that requires verbal skills in English can be daunting. All other things being equal, over-representation in science and engineering should be expected for a language-based minority. But for Hispanics attrition rates in science and engineering were also unusually high.

Similar results were obtained by Rogers Elliott and his co-investigators. In their study, they found that 70 percent of Asians persisted in their ambition, while 61 percent of whites, 55 percent of Hispanics and 34 percent of blacks did.[72] Others had similar findings.[73] Again, I am aware of no contrary evidence.

### *C. Students with Low Entering Credentials in Science, Both in Absolute and in Comparative Terms, Are More Likely to Leave Science & Engineering*

It is tempting to ask the question "What accounts for disproportionate minority attrition?" first. But that temptation should be avoided. Instead, the first question should be "What accounts for student attrition in general?" Once that preliminary question is answered, the question about disproportionate minority attrition essentially answers itself.

It is no secret that entering science credentials–like Math SAT score and the number and grades received for high school courses in mathematics and science–are strongly correlated with persistence in science.[74] Since African-American, Hispanic and American Indian students tend as a group to have lower entering science credentials, they are almost certain to have a higher attrition rate from science and engineering.[75]

It would be nice if the disparities among the races, including the disparities between Asians and others, could be eliminated overnight by improving the performance of underrepresented minorities.[76] For that matter, it would be nice if disparities between individuals could be eliminated and everyone could perform better in mathematics, science, and all subjects.

There is no doubt that improvements can be made even within the constraints of the current system. Dr. Thomas Fortmann testified before the Commission, for example, that more needs to be done to ensure that mathematics and science classes at the elementary and secondary levels are being taught by qualified teachers. The Commission has responded by recommending that K-12 schools "recruit qualified math and science teachers using, if necessary, pay adjustments and incentives."[77]

But if there is one thing that we have learned during the many decades that this problem has been receiving attention, it is that few improvements can be made quickly. The mismatch problem, however, may be a partial exception. Matching students to the right college or university for their level of developed academic ability could increase the number of science and engineering majors in fairly short order.

As three independent studies have now concluded, absolute credentials are not the only thing that matters in keeping college students in science and engineering. Relative credentials are also important. A student whose entering credentials are at the bottom of the class at the school he attends is less likely to persevere in his quest for a degree in mathematics or engineering than a student with identical credentials who attends a school where those credentials place him higher in the class.

The first of these studies was that published by Rogers Elliott and his co-investigators in 1996.[78] The single most important culprit they found was the "*relatively* low preparation of black aspirants to science in these schools."[79] The Elliott team was careful to put the emphasis on "relatively." It wasn't just entering credentials demonstrating highly developed ability at science that mattered, but comparatively high credentials. A student who attended a school at which his Math SAT score was in the top third of his class was more likely to follow through with an ambition to earn a degree in science or engineering than was a student with the same score who attended a school at which his score was in the bottom third. The following chart was presented:

According to the authors, the bottom line was this: A student with an SAT Math score of 580 "who wants to be in science will be three or four times more likely to persist at institutions J and K, where he or she is competitive, than at institutions A and B, where he or she is not."[81]

For some this is counter-intuitive. The more prestigious the school, they believe, the more adept it should be at graduating future physicians, scientists, and engineers, no matter what their entering credentials. But instructors everywhere must pitch the material they teach at a particular level. They can pitch to the top of the class, to the middle or to the bottom, but they can't do all three at the same time. At elite colleges and universities pitching to the bottom of the class is uncommon–especially in the science and engineering departments. The whole point of these institutions is to teach to the top. That is the reason that students, who may have been positively mismatched in high school, are willing to travel thousands of miles and incur significant debt to attend them. If they were to abandon that practice and resolve to teach to the bottom of the class, they would no longer be elite institutions.[82]

**Table II. Percentage ofEarned Degrees in the Natural Sciences as a Function of Terciles of the SAT-MDistribution in 11 Institutions[80]**

| | Tercile 1 | | Tercile 2 | | Tercile 3 | |
|---|---|---|---|---|---|---|
| Institution | % Degrees | SAT-M | % Degrees | SAT-M | % Degrees | SAT-M |
| Institution A | 53.4 | 753 | 31.2 | 674 | 15.4 | 581 |
| Institution B | 57.3 | 729 | 29.8 | 656 | 12.9 | 546 |
| Institution C | 45.6 | 697 | 34.7 | 631 | 19.7 | 547 |
| Institution D | 53.6 | 697 | 31.4 | 626 | 15.0 | 534 |
| Institution E | 51.0 | 696 | 34.7 | 624 | 14.4 | 534 |
| Institution F | 57.3 | 688 | 24.0 | 601 | 18.8 | 494 |
| Institution G | 62.1 | 678 | 22.6 | 583 | 15.4 | 485 |
| Institution H | 49.0 | 663 | 32.4 | 573 | 18.6 | 492 |
| Institution I | 51.8 | 633 | 27.3 | 551 | 20.8 | 479 |
| Institution J | 54.9 | 591 | 33.9 | 514 | 11.2 | 431 |
| Institution K | 55.0 | 569 | 27.1 | 472 | 17.8 | 407 |
| Medians | 53.6 | | 31.4 | | 15.4 | |

The extraordinary record of Historically Black Colleges and Universities was one source of evidence cited by the Elliott team in favor of their conclusion. With only 20 percent of total African-American enrollment, these schools produce 40 percent of the African-American students graduating with natural science degrees according to the National Science Foundation. These students frequently go on to earn Ph.D.s from mainstream universities. The National Science Foundation reports, for example, that of the approximately 700 African-Americans who earned a doctorate in science or engineering between 1986 and 1988, 29 percent earned their undergraduate degree from an HBCU. For biologists, the figure was 42 percent and for engineers it is was 36 percent.[83] Even those who have mixed feelings about HBCUs (and I am such a person) must admit this is impressive.

Why have HBCUs been so successful? Unlike at mainstream institutions with their high levels of affirmative action, African-American students at HBCUs are not grouped at the bottom of the class. Roughly half of African-American students at HBCUs will be in the top half of the class. Many will be honor students. As a result, systematic mismatch is just not an issue.[84]

The problem is not that there are no minority students capable of doing honors work at mainstream college and universities. There are many. But there are not enough at the very top tier to satisfy the demand for diversity. And when elite universities like Cal Tech, MIT or the Ivies lower their academic standards in order to admit a more racially diverse class, schools one or two tiers down feel they must do likewise, since the minority students who might have attended those schools based on their own academic record are instead attending the more elite schools. The problem thus cascades downward to the fourth and fifth tiers, which respond similarly. As a result, a serious gap in academic credentials between minority and non-minority students is created at all competitive levels at mainstream universities–a gap that results in seriously disappointing grades for many minority students, especially in science and engineering classes where good grades are hard to come by.

At least one HBCU faculty member–Professor Walter Pattillo, Jr. of North Carolina Central University–intuitively grasped the mismatch problem even before the Elliott team was able to demonstrate its existence empirically. As then-chairman of the biology department, he vented his frustrations to *Science* magazine in 1992: "The way we see it, the majority schools are wasting large numbers of good students. They have black students with admission statistics [that are] very high, tops. But these students wind up majoring in sociology or recreation or get wiped out altogether."[85]

Neither Professor Pattillo nor the Elliott study received attention from mainstream college or university administrators. Admissions policies at competitive schools continued to emphasize the importance of recruiting minority students even if their entering credentials would put them towards the bottom of the class.[86] Instead, emboldened by their perception that the Supreme Court had given a constitutional green light to racially preferential admissions policies in *Grutter v. Bollinger* (2003),[87] selective college and universities ramped up those policies.[88] The supposed beneficiaries of these policies–minority members aspiring to become physicians, engineers and scientists–were not informed.

Around that time, however, the tide of opinion among social scientists studying the issue was beginning to turn, even as is remained frozen among college and university administrators.[89] One of the milestones was the publication of *Increasing Faculty Diversity: The Occupational Choices ofHigh Achieving Minority Students* in 2003. The long-term project was funded by the Mellon Foundation, which had been and remains one of the nation's most zealous institutional backers of race-based admissions policies. The authors' mission was to determine why more minority students are not attracted to careers in academia. Their conclusions, reached after extensively questioning 7,612 high-achieving undergraduates at 34 colleges and universities, pointed to mismatch as a significant culprit:

> The best-prepared African-Americans, those with the highest SAT scores, are most likely to attend elite schools, especially at the Ivy League. Because of affirmative action, these African-Americans (those with the highest scores on the SAT) are admitted to schools where, on average, white students' scores are substantially higher, exceeding those of African-Americans by about 200 points or more. Not surprisingly, in this kind of competitive situation, African-Americans get relatively low grades. It is a fact that in virtually all selective schools (colleges, law schools, medical schools, etc.) where racial preferences in admission is practiced, the majority of African-American students end up in the lower quarter of the class....
>
> African-American students at the elite schools (the liberal arts colleges and the Ivy League) get lower grades than students with similar levels of academic preparation (as measured by SAT scores) than African-American students at the nonelite schools (state universities and HBCUs). ***Lower grades lead to lower levels of academic self-confidence, which in turn influence the extent to which African-American students will persist with a freshman interest in academia as a career. African-American students at elite schools are significantly less likely to persist with an interest in academia than are their counterparts at nonelite schools***.[90]

To say that the Mellon Foundation was not happy with the conclusions of its grant recipients would be an understatement. Soon after publication, the *Chronicle ofHigher Education* reported that the foundation was "trying to distance itself" from the book's findings.[91] Unlike similar projects with Mellon funding, this one did not receive a publicity push from the foundation.

Dr. Cole told the *Chronicle* that there was "no chance" that he would receive money again from the Mellon Foundation. "And I don't care," he said. "I was trained at a time before social science became so politicized." "I believe that social science should be objective and value-free, and you should design a study to answer a question and whatever the answer is, that's what it is."[92]

A year after Cole & Barber's research became public, a second study on the science and engineering mismatch issue was published. University of Virginia psychologists Frederick L Smyth and John J. McArdle used a very different methodology and a different database from those of Elliott and his co-authors. But they reported findings that "are consistent" with the earlier article's conclusion that "race-sensitive admissions, while increasing access to elite colleges, was inadvertently causing disproportionate loss of talented underrepresented minority students from science majors."[93]

Indeed, Smyth & McArdle went further. They developed a model that attempts to measure how many more minority students would have succeeded in the goal of a science or engineering degree if colleges and universities had employed race neutral admissions criteria. They wrote:

> "According to our model ..., if all the [Science-Mathematics-Engineering]- intending underrepresented minority students had enrolled in similarly functioning colleges where their high school grades and math test scores averaged at the institutional means among

> [Science-Mathematics-Engineering] intenders, 72 more of the women and 62 more of the men would be predicted to persist in [Science-Mathematics-Engineering] (45 percent and 35 percent increases, respectively)."[94]

Smyth & McArdle's recommendation was very clear: "Admission officials are advised to carefully consider the relative academic preparedness of science-interested students, and such students choosing among colleges are advised to compare their academic qualifications to those of successful science students at each institution." Again, few college or university administrators were listening.

The latest contribution to the literature on mismatch in science and engineering is Do Credential Gaps in College Reduce the Number of Minority Science Graduates?[95] Using a number of sophisticated methodologies, Sander & Bolus arrive at conclusions like those of Smyth & McArdle and the Elliott team.

Sander & Bolus studied data obtained from the multi-campus University of California. All campuses of the University of California are quite selective. But some are more selective than others. The flagship campus at Berkeley is highly selective as are the UCLA campus and the more science-oriented UC-San Diego. At the other end of the spectrum, the campuses at Riverside and Santa Cruz are easier to gain admittance to, but nonetheless hardly "easy."

Employing what they call the "distance method," Smyth & Bolus measured the distance between each student's entering academic index and the median academic index of all science and engineering-interested students at that campus. This allowed the authors not just to compare students with equal academic indices attending different University of California campuses, but also to make comparisons based on the magnitude of mismatch. Since the UC campuses differ in their median academic index, students with equal academic indices but attending different campuses will differ in their level of mismatch.[96]

They found that students who are "mismatched" at one University of California campus are at a greater risk of failing to attain their initial goal of a science or engineering degree than otherwise identically-credentialed students attending a less selective campus of that same university at which they were not mismatched. And the greater the mismatch, the greater the problem.

Not satisfied with confining their analysis to the "distance method," Sander & Bolus also employed what they dubbed the "first choice/second choice" method. This approach involves looking at pairs of students who were admitted to two different UC campuses, one more elite and the other less elite. In each pair, one student chose to attend the more elite school and the other the less elite. The results were the same: Mismatched students are at a disadvantage in science and engineering.[97]

"Minority attrition in science is a very real problem, and the evidence in this paper suggests that 'negative mismatch' probably plays a role in it," they wrote. The approaches they took yielded consistent results: "[S]tudents with credentials more than one standard deviation below their science peers at college are about half as likely to end up with science bachelor degrees, compared with similar students attending schools where their credentials are much closer to, or above, the mean credentials of their peers."[98]

### *D. Conclusion*

Decades ago, well-meaning administrators at selective college and universities resolved to "do the right thing" by extending preferential treatment to underrepresented minorities in admissions. One of the consequences of that policy has been systematically low college grades for the supposed beneficiaries of that preferential treatment.[99] No, it doesn't apply to all such students, but it is nevertheless a widespread phenomenon. And the reason is simple: While some students will outperform their entering academic credentials, just as some students will underperform theirs, most students will perform in the range that their entering credentials predict.

No serious supporter of affirmative action denies this. William G. Bowen and Derek Bok, authors of The Shape of the River: Long-Term Consequences of Considering Race in College and University Admissions and long-time advocates of race-based admissions policies, candidly admit that the credentials gap has serious consequences: "College grades [for affirmative action beneficiaries] present a ... sobering picture," they wrote. "The grades earned by African-American students at the [schools we studied] often reflect their struggles to succeed academically in highly competitive academic settings."[100]

The long-term social and educational consequences of decades of race-based admissions policies and the artificially low grades for minorities those policies produce are only now beginning to be studied. The evidence examined by the Commission on Civil Rights focuses only on the effects on science and engineering majors. It suggests that, as a result of race-based admissions policies, we now have fewer, not more, physicians, dentists, engineers, scientists and other science-oriented professionals than we would have had under a policy of color-blindness.

While there are still a few unanswered questions, it is time for students to be advised of the issue and allowed to make their own decision about their future. Indeed, it is long past time. If higher education were held to the same standards of consumer disclosure as other businesses–from securities brokerage houses to children's toy manufacturers–disclosure would have come long ago.

## Statement of Commissioner Ashley L. Taylor, Jr.

It is my sincere hope that readers of this report picked up on a significant yet subtle message in the second paragraph of the Executive Summary clarifying that this topic was limited to the acceptance of minority students to very rigorous academic programs at elite, selective institutions where the institution is very unlikely to provide any remedial assistance to overcome K-12 academic deficiencies. This report is not intended to address the admissions or success of all minority students who pursue a science, technology engineering or math (STEM) course of study at all universities. It is a narrow look at the fate of the students who find themselves selected by an institution into a course of study for which they are underprepared or have little support.

The STEM disciplines are difficult for all students, but especially for students whose K-12 experience, both in the school and at home, left them lacking critical academic skills or critical study skills. Unlike the humanities, arts, literature, even business courses of study, the STEM disciplines build quickly from one level to the next leaving little time or room to play "catch up" for a student who starts the program at a disadvantage. Unfortunately, there remains a persistent gap between non-Asian minorities and whites in K-12 preparation insofar as it relates to the profile of students admitted to elite institutions. It is that small but critically important high performing minority slice of the student population which needs information that can assist them in choosing the right college to attain their future goal.

I particularly support the first two recommendations of this report from a consumer protection standpoint. Minority students who are applying to or being recruited by elite institutions for STEM programs need to know certain fundamental realities. There is a delicate balance between full disclosure and outright discouragement of students' ambitions. But, there is a higher price paid by the student (and the student's family) who drops out of a STEM program than is paid by the institution, and for that reason I prefer to err on the side of full disclosure.

That said, I believe that every student should be viewed as an individual. We should never place people in typecast such that we eliminate their opportunity to expand and excel. Students should be given every opportunity to do so, and we shouldn't make assumptions about any individual. That is the beauty of our country -- we don't do that. We don't have scripts for people depending on what class of society they come from. It is also important, however, that the student be made aware of the facts. Those facts could be brutal, those facts could be demoralizing and I understand the reluctance and hesitation of presenting those facts to an 18 year-old young adult who may not have the capacity to properly process and assimilate that information.

The schools have a duty to provide the support for every student they admit. Unfortunately, I believe that the schools focus a lot more on admissions and matriculation and not enough on graduation and success upon graduation. When the school doesn't uphold its end of the bargain, students doesn't know how much force they are going to have to apply to the situation or how much harder they are going to have to work than the person sitting beside them. I believe that the students we are talking about in this report would apply that pressure if they knew it was needed.

In life, the only way you can know how much pressure to apply to a situation is if you know what you are up against. I am very concerned that some minority students are walking into certain academic environments where the school fails to provide that support naturally. To those students who are in those difficult situations and are aware of the uphill climb, I want to encourage them. I want them to know there are people and resources that can help them if they seek it.

On a final note, I have intentionally elected not to use the term "mismatch" which was presented during the briefing and is used in the report. The word, in my view, carries connotations which may be easily misconstrued.

## Dissent of Commissioners Michael Yaki and Arlan D. Melendez

We strongly dissent from this Report and most of its Findings and Recommendations for several reasons. Specifically, we dissent from: 1) Part A; 2) Findings 2, 3, 4, 5, 6, and 7; and 3) Recommendations 1, 2, 3, and 4. Our principal objection to this Briefing and Report is that they were fundamentally not about encouraging minorities to pursue careers in STEM fields. Rather, the major focus of the briefing and report was to promote Rogers Elliott, Richard Sander and their "mismatch" theory.

As was noted at the Briefing Hearing, the panel was unbalanced and stacked in favor of "mismatch" proponents.[101] Even Mr. Elliott conceded that his comments were substantially similar to those of Mr. Sander.[102] In addition to being mostly a reaffirmation of Mr. Sander's conclusions, Mr. Elliott's testimony was based on his paper which was already well over a decade old by the time of our Briefing Hearing.

Mr. Sander's efforts to promote the theory of "mismatch" with regard to law school and the legal profession[103] have been soundly debunked in a sustained manner.[104] Mr. Sander's attempts to address his critics and salvage his initial claims[105] have resulted only in further critiques that have supported the initial critiques and have pointed out that Sander's self-defense has resulted in self-undermining.[106] Perhaps in response to the discrediting of his initial claims concerning the effect of "mismatch," Sander himself has avoided using the term "mismatch" in his most recent work.[107]

The Report's recommendations are largely premised on claims regarding the effect of so-called "mismatch." This effect has been exaggerated by both Elliot/Sander and their admirers on the Commission.[108] We are concerned that the recommendations, grounded as they are in these out-sized claims, would have the effect of discouraging minorities from pursuing careers in STEM fields. This outcome would be diametrically opposed to the stated purpose of this report.

Additionally, we object to the non-deliberative manner by which the Findings and Recommendations were drafted and adopted. As has been become the norm with the Commission, Commissioner-drafted alternatives to staff-issued findings and recommendations are only being provided to all Commissioners less than a day before votes are scheduled on the findings and recommendations. In some cases, alternate findings and recommendations have been issued the night before a Commission meeting, only to be superseded by even newer drafts issued the morning of a meeting or during the meeting itself.

## Joint Rebuttal of Commissioners Gail Heriot, Peter Kirsanow and Todd Gaziano

Commissioners Yaki and Melendez take the position that the 1996 article by Dartmouth College psychology professor Rogers Elliott and his co-authors is "over a decade old" and hence in their view too old to be valuable.[109] Perhaps they regard the 2009 article by UCLA law professor Richard Sander and UCLA Medical School senior statistician Roger Bolus as tainted because it is somehow too new.[110] But, if so, that would still leave the 2004 article by University of Virginia psychology professors Frederick L. Smyth and John J. McArdle.[111] Not all these empirical studies can be dismissed so casually.

All three reach the same conclusion: Mismatch likely puts affirmative action beneficiaries who plan to major in science or engineering at special disadvantage. More specifically, all three agree that a student whose entering credentials put him towards the bottom of his college class is less likely to follow through with plans to major in science or engineering than a student *with the same entering credentials* attending a less elite school where those credentials put him in the middle or towards the top of the class.[112]

Is it possible that Drs. Elliott, Strenta, Adair, Matier, Scott, Smyth, McArdle, Sander, and Bolus are nevertheless incorrect? Of course it's possible. That is why more research would be useful. At this point, however, the evidence is going strongly in favor of their conclusions. And that is important to recognize.

The Commission has nevertheless taken a cautious approach. It is not calling upon Congress or state legislatures to prohibit racially-preferential admissions policies for students who intend to major in science or engineering. Nor is it calling on colleges and universities to abandon racial preferences altogether, at least not in this report. The Commission's recommendation is modest. What we are asking is that college applicants be made aware of the potential for a mismatch effect. Students should be given the facts, so that they, in consultation with their parents, teachers and other advisors, can decide for themselves how to proceed.[113]

It is difficult to understand why Commissioners Yaki and Melendez object to disclosure. Next to homeownership, a college education is typically the largest investment an American makes during his or her lifetime. The law requires that manufacturers of breakfast cereals inform consumers of the ingredients and nutritional content of their products as well as warn of any known health hazards associated with them. Here we are simply calling for disclosures by the institutions themselves. Is the value of a college degree less important than corn flakes?

The dissenters could argue that the Commission should wait until all the evidence is in before they make any recommendation, even a recommendation of disclosure.[114] But in issues of public policy, the evidence is never all in. Decisions must be made based on the available evidence, not the evidence that would exist in a perfect world. At present, the evidence that students intending to major in science and engineering can be worse off if they are mismatched is fairly strong—much stronger than the evidence that underlies most public policy decisions. Indeed, it is considerably stronger than the evidence that racially-preferential admissions policies benefit minority students in any context, science-related or not.[115]

Commissioners Yaki and Melendez argue that the panel at the briefing was "stacked in favor of 'mismatch' proponents."[116] Since no researcher has taken a position contrary to that of our witnesses, Dr. Elliott and Dr. Sander, on the issue of mismatch in science and engineering, it is difficult to understand who Commissioner Yaki believes should have been called as a witness. The Commission cannot call witnesses that do not exist.

Nonetheless, extraordinary efforts were made to call either (1) any witness who had expressed doubts about the problem of mismatch in any context or (2) any witness who had expressed strong support for race-based admissions as a means of increasing the number of minority members in science & engineering.[117] Among those contacted were: Gibor Basri, Vice Chancellor for Equity and Inclusion at the University of California at Berkeley; Ted Greenwood, Program Director at the Alfred P. Sloan Foundation; Freeman A. Hrabowski III, President of the University of Maryland (Baltimore County); Christopher Jencks, Professor of Social Policy at Harvard University; Paul Joskow, President of the Alfred P. Sloan Foundation; Thomas Kane, Professor of Education at Harvard University; William Kidder, Special Assistant to the Vice President for Student Affairs, University of California; Tom Loveless, Senior Fellow in Governance Studies at the Brookings Institute, Tom Luce, CEO of the National Science and Math Initiative; and Kenneth Maton, Professor of Psychology at the University of Maryland (Baltimore County).

Commission staff members also invited Ann Mullen, Associate Professor of Sociology at the University of Toronto; Meredith Phillips, Associate Professor of Public Policy & Sociology at UCLA, Jesse Rothstein, Associate Professor of Public Policy at Princeton University; Marta Tienda, Professor of Sociology & Public Affairs at Princeton University;

Philip Uri Treisman, Professor of Mathematics & Public Affairs at the University of Texas; and Sarah Turner, Professor of Economics & Education at the University of Virginia.[118]

All declined to testify. Of course, if Commissioners Yaki and Melendez could have suggested additional potential witnesses. Indeed, they were asked to by the staff--in writing and more than once. But they did not respond with recommendations—even though there is no question that any such recommendations would have been followed.

The peculiar thing about the complaint of the dissenting commissioners is that ultimately the panel was fairly well-balanced. Commission staff members were able to identify and secure the testimony of a qualified witness, Rice University Professor of Mathematics Richard Tapia, a long-time forceful advocate of increasing the number of minority students in science and engineering at highly competitive institutions.[119] Still, Commissioner Yaki at least appeared disappointed rather than pleased and complained of lack of balance anyway. His reaction when Fisk University President Hazel O'Leary (former Clinton Administration Secretary of Energy) abruptly cancelled her testimony on the eve of the briefing (not long after the identities of the witnesses were announced to the Commissioners) was equally if not more telling: Commissioner Yaki declared that he was "kind of glad she did" and implied that the Commission's calling of President O'Leary to testify on the successes of HBCUs in graduating African-American science and engineering majors was somehow inappropriate.[120]

Finally, the objections of Commissioners Yaki & Melendez to what they regard as the "non-deliberative manner by which Findings and Recommendations were drafted" are surprising. In fact, what they are objecting to is the deliberative character of our negotiations rather than their non-deliberative character. The Commission members who make up the majority that adopted these findings and recommendations take them seriously. We are not potted plants. We do not show up once a month to sign on the dotted line where staff members tell us to, and then collect a stipend and a free lunch. We sometimes have been negotiating with each other over the language of the findings and recommendations in a report long prior to their adoption. But given that all of us have full-time jobs and most of us must travel distances to get to our meetings, these negotiations take a while. Sometimes changes are still being suggested both shortly before a meeting and during the meeting. If Commissioners Yaki and Melendez would like to take part in those negotiations, they would certainly be welcome to, but thus far they have shown no desire to do so. To the contrary, Commissioner Yaki has declared the opposite intention in an open meeting, "I will not be a party," he said to Chairman Reynolds, "to contributing to your ability to get things done."[121] In any event, the Chairman has always been willing to postpone consideration of findings and recommendations when Commissioners Yaki or Melendez have requested such a delay, so it is difficult to understand what motivates their argument.

Part of the reason Commission members must spend a good deal of time hammering out findings and recommendations lies in the Commission's ill-considered procedures. The staff members who are assigned to write the first draft of them seldom get as much time as I would like to confer with Commission members about the kinds of recommendations those Commission members want to make. In part this is a result of poorly-designed internal procedures, which deliberately isolate them from the Commission. Sometimes staff members end up making guesses about what the Commission's majority wants, and sometimes those guesses turn out to be incorrect. They learn. Ultimately, however, it is for the Commissioners and not for the staff members to issue the reports. That is what it means to be the Commissioners instead of the staff. As the statute that chartered the Commission makes clear, the Commission is made up of the eight Commissioners; it is not made up of the staff.[122] The findings and recommendations of Commission reports must reflect that.

## End Notes

[1] The terms "minority," "non-Asian minority," and "underrepresented minority" refer to the same group and were used interchangeably by the panelists.

[2] STEM Briefing Tr. (hereinafter Tr.) at 11-23. Professor Sander stated that he was presenting preliminary research, subject to revision, based on data obtained in July 2008 from the University of California, Office of the President. Dr. Sander stated that he was collaborating with several other scholars in a study of science mismatch in the University of California system that he hoped would be completed in 2009. During his testimony, he referred to projected images (slides 1–20) which can be found at the end of this summary.

[3] Please refer to slide 2; slide 1 is a title page. Subsequent references to slides will appear in the text.

[4] His data interpretation showed relative proportions of general and science degrees earned by college-aged students of different races compared to a white norm set at 100. This data interpretation showed that black students were 56 percent as likely to earn a BA of any kind, and 43 percent as likely to earn a Ph.D.

[5] According to Dr. Sander, "positive mismatch" occurs when a student is not challenged, since the student can learn more and at a faster rate than his classmates; "negative mismatch" occurs when a student falls behind or feels overwhelmed because of the speed or complexity with which material is presented.

[6] William Bowen & Derek Bok, *The Shape of the River*, (Princeton University Press 2000).

[7] Bowen and Bok looked only at elite schools and only at SAT scores, not at other academic credentials like high school grades. At some schools, students with SAT scores above the school's median may tend to have lower high school grades than the average student's and vice versa; indeed this may account for why the student is attending that school and not a more (or less) competitive one. As a consequence, efforts to show how well SAT scores predict academic success that do not control for high school grades will obscure rather than reveal an important relationship.

[8] Dr. Sander noted that in slide 10's data set, blacks (and to a lesser extent, Hispanics) are far more clustered in the lower-index/high-preference ranges at Michigan than are whites and Asians, meaning that conclusions as to high-preference white rates are based on relatively few data points.

[9] *See* http://www.ssrc.org/press/cla-study (accessed Apr. 6, 2009).

[10] Tr. at 23-24.

[11] *See* Dr. Tapia's statement in the "Statements" section.

[12] In his written statement, Dr. Tapia stated: "The mismatch theorists have focused light on a huge problem that I have been fighting at Rice for 20 years. The data they present are entirely credible to me because they reflect what I have seen at Rice. So I do not dispute the data. It is the recommendations they make based upon the data that are terribly flawed." See Tapia Statement below at page 35 of this report.

[13] Tr. at 28.

[14] Tr. at 34.

[15] *See* http://www.caam.rice.edu/~rat/cv/minority/mismatch_theory_statement_2008-09-12.htm (accessed Jan. 8, 2009).

[16] Tr. at 34–43.

[17] Rogers Elliott, A. Christopher Strenta, Russell Adair, Michael Matier and Jannah Scott, "The Role of Ethnicity in Choosing and Leaving Science in Highly Selective Institutions," 37 Res. in Higher Ed. No. 6 at 681-709 (1996).

[18] Dartmouth College; Brown, Cornell, and Yale universities. Tr. at 39.

[19] Students who dropped out of science programs did not usually drop out of school; they may have switched to other majors.

[20] Dr. Elliott's data as presented to the Commission in three slides, showed an entering average Hispanic SAT math score of 653 and an entering average score for black students of 607. The slides are reproduced immediately after this summary.

[21] For a more detailed explanation, please see the attached statement of Dr. Elliott, also available at http://www.seaphe.org/pdf/elliott-ethnicity.pdf (accessed Jan. 8, 2009).

[22] Dr. Elliott submitted corrections by communications dated April 1, 2009 and July 7, 2009. SAT score data for 2008 show that black students who scored at or above 650 on the SAT math section were at the 98th percentile of black SAT-takers that year. *See* http:// professionals.collegeboard.com/profdownload/sat_ percentile_ranks_ 2008_cr_m_w_gender_ethnic_group s.pdf (accessed Feb. 20, 2009).

[23] Dr. Elliott referred during his testimony to three slides on view during his briefing testimony. This is labeled "Slide 3" in this Commission report. This table can also be found at page 701 of that article and is entitled, "Table 4. Percentage of Earned Degrees in the Natural Sciences as a Function of Terciles of the SATM Distribution in 11 Institutions."

[24] Dr. Elliott noted that the biggest undergraduate science degree-granting schools for blacks in this country were not considered prestigious.

[25] Tr. at 43-52.

[26] *See* http://www.massachusetts.edu/stem (accessed Apr. 3, 2009).

[27] Tr. at 49.

[28] Tr. at 47.

[29] Tr. at 52-62.

[30] Tr. at 63-149.

[31] *See*, e.g., Roland Fryer, Jr. and Steven Levitt, "The Black-White Test Score Gap Through Third Grade," NBER Working Paper Series, Working Paper 11049, National Bureau of Economic Research, http://pricetheory.uchicago.edu/levitt/Papers/FryerLevitt2005.pdf (accessed on September 21, 2009); 8 Am. Law Econ. Rev. 249-281 (2006).

[32] Dr. Fortmann cited a case study of 15 schools in "It's Being Done: Academic Success in Unexpected Schools," by Karin Chenoweth, published by Harvard Education Press (2007).

[33] "Rising Above the Gathering Storm," http://www.nap.edu/catalog.php?record_id=11463 (accessed January 13, 2009).

[34] *See Grutter v. Bollinger*, 539 U.S. 306 (2003).

[35] Peter Arcidiacono, Shakeeb Khan, and Jacob Vigdor, "Representation versus Assimilation: How do Preferences in College Admissions Affect Social Interactions?" November 5, 2007, available at http://www. econ.duke.edu/~psarcidi/knewminisk.pdf (accessed Jan. 23, 2009).

[36] Tr. at 111–12.

[37] Princeton's current average SAT verbal and math combined score is 1485. *See* http://collegesearch.collegeboard. com/search/CollegeDetail.jsp?collegeId=4221&profileId=6 (accessed Feb. 24, 2009).

[38] One definition is as follows: "Critical Race Theory (CRT) is an intellectual movement of progressive law scholars–primarily of color–who view the law as complicitous in sustaining white supremacy, and, by extension, upholding similar hierarchies within gender, class, and sexual orientation." http://tarlton. law.utexas.edu/lpop/etext/lsf/isaksen24.htm (accessed September 17, 2009).

[39] This represents about a one standard deviation gap in the LSAT.

[40] Elliott, Rogers, et al. "The Role of Ethnicity in Choosing and Leaving Science in Highly Selective Universities." *Research in Higher Education.* Vol. 37, No. 6 (1996). http://www.springerlink.com/content/ 1662845 727427n33/

[41] Sander, Richard, 2005. "A Systemic Analysis of Affirmative Action in American Law Schools," *Stanford Law Review* 57 (2) 367-484.

[42] Tapia, Richard A. "True Diversity Doesn't Come From Abroad." Chronicle of Higher Education, p. B34. September 28, 2007.

[43] Created to avoid the impact of *Hopwood v. Texas*, the case banning the use of race as a factor in admissions, 1997's Texas House Bill 588, guarantees Texas students who graduate in the top ten percent of their high school class automatic admission to all state-funded universities.

[44] Steven Pinker, *How the Mind Works* (Norton, 1977), p. 341

[45] LIPING MA, KNOWING AND TEACHING ELEMENTARY MATHEMATICS (Lawrence Erlbaum Associates, 1999). For a review and summary see www.aft.org/pubs-reports/american_educator/ fall99/amed1.pdf.

[46] NATIONAL COUNCIL ON TEACHER QUALITY, NO COMMON DENOMINATOR: THE PREPARATION OF ELEMENTARY TEACHERS IN MATHEMATICS BY AMERICA'S EDUCATION SCHOOLS (June 2008): www.nctq.org/p/publications/reports.jsp.

[47] NATIONAL MATHEMATICS ADVISORY PANEL, FOUNDATIONS FOR SUCCESS (Final Report, March 2008): www.ed.gov/about/bdscomm/list/mathpanel/report/final-report.pdf.

[48] Details are in the Commissioner's Guidelines for the Mathematical Preparation of Elementary Teachers (July 2007): www.doe.mass.edu/mtel/MathGuidance.pdf

[49] Robert Moses, voting-rights pioneer and founder of the Algebra Project, argues in his marvelous book RADICAL EQUATIONS (Beacon Press, 2001) that math literacy is a civil right.

[50] Apart from the considerations discussed in this report, there may indeed be something special about the education available from America's most academically competitive colleges and universities. It should be noted, however, that some of the most sophisticated research available suggests that when it comes to increasing one's income, elite schools are not exactly the ticket. See Stacy Berg Dale & Alan B. Krueger, Estimating the Payoff to a Highly Selective College, 117 Quarterly J. Econ. 1491 (2002). Graduates of Ivy League institutions are indeed high earners. But, if this research is correct, this is simply a reflection of the fact that very talented students attend those schools. If the same students had attended less prestigious schools, they would have done on average just as well financially–or so this research suggests.

In any event, the competition for admission to elite colleges and universities is not a bad thing in itself. To the contrary, students who work hard to hone their skills in order to get into their first-choice school have improved their skills even if ultimately they are not admitted to that school. But occasionally the competition can be over-the-top. Some parents, for example, are willing to fork over as much as $40,000 for advice from so-called experts in getting their son or daughter into an Ivy League school. Susan Berfield & Anne Tergeson, I Can Get Your Kid into an Ivy: Michele Hernandez Boasts that 95 percent of Her Teenage Clients are Accepted by Their First-Choice School. Her Price: As Much as $40,000 a Student, Business Week (October 22, 2007).

[51] In general, when I refer to science and engineering in this statement, I mean to include science, technology, engineering and mathematics, which are collectively referred to as "STEM" in the summary of the proceedings. The exception is when I discuss the findings of individual empirical studies. There I use terms as they are used in the particular study being discussed.

[52] Dana Milbank, Education: Shortage of Scientists Approaches a Crisis As More Students Drop Out of the Field, Wall Street Journal (September 17, 1990).

[53] *See* infra at Part C.

[54] See infra at Part C.

[55] As one early researcher on this topic put it, there is an academic advantage to being a "big frog" in the "frog pond." James A. Davis, The Campus as a Frog Pond: An Application of the Theory of Relative Deprivation to Career Decisions of College Men, 72 Am. J. Socio. 17 (July 1966). This article was written well before the concept of mismatch came to be associated with the controversy over affirmative action and was not focused specifically on science and engineering. Davis found that college GPA is more strongly correlated with career choice than is quality of institution. In other words, while students take both their college grades and the academic quality of the school they are attending into account in evaluating their career choices, they tend to place more emphasis on college grades than is justified. A student at the bottom of his very elite class will discount his abilities more than is called for, while a student at the top of a class at a mediocre school will overestimate his ability. Davis concludes: "[T]hese ideas have some implications for educational policy. At the level of the individual, they challenge the notion that getting into the 'best possible' school is the most efficient route to occupational mobility. Counselors and parents might well consider the drawbacks as well as the advantages of sending a boy to a 'fine' college, if, when doing so, it is fairly certain he will end up in the bottom ranks of his graduating class." Id. at 30-31.

[56] While the observation that a student is more likely to learn in a classroom where his academic credentials are not at the bottom of the class is common sense, it is not necessarily true in all contexts. Take, for example, a school with 100 first graders. If one were to divide the class into thirds according to their achievement test scores, it is not necessarily the case that all three groups would learn more than if they were divided into three groups at random. Students with behavioral problems may tend to be over-represented in the group of students with the lowest scores for the simple reason that their behavior has interfered with their own learning. Concentrating them in one group may create havoc in the classroom that interferes with the ability of all students, not just those with behavioral problems, to learn.

[57] *See*, e.g., Thomas Sowell, Inside American Education: The Decline, the Deception, the Dogmas (1993).

[58] Mismatch may be positive or negative. If a typical Cal Tech freshman were to take a Basic Physics class designed for law professors like me, many of whom have never excelled at science, she would likely learn less than she would have in a class with her fellow Cal Tech students. Coasting through a "Basic Physics for Dilettantes" course, she would be the victim of positive mismatch, while I am negatively mismatched in the hypothetical.

[59] The empirical studies discussed in this statement, see infra at Part C, do not distinguish among the reasons that mismatched students might drop out of science and engineering more often than non-mismatched students with similar credentials. They simply record that they disproportionately do so. Is it just because they perceive that they aren't doing well relative to other students and hence lack confidence in themselves? Or are they actually learning less than their similarly-credentialed counterparts who persevere in science or engineering at somewhat less elite institutions? Or both? There is, at present, no national examination for science and engineering achievement that would allow researchers to determine whether college students who were mismatched and dropped out of science or engineering actually learned less than their counterparts at less elite schools who took similar courses. The intuitive answer is that they did and that their self-confidence was also shaken in the process. But it is unnecessary at this point to draw a distinction. The law school experience is clearer, since law students must pass a bar examination in order to practice law. There is empirical evidence that mismatched law students are less likely to pass the bar examination than their non-mismatched counterparts at less elite schools. Richard Sander, A Systemic Analysis of Affirmative Action in American Law Schools, 57 Stan. L. Rev. 367, 393 (2004).

[60] While the Supreme Court case of *Gratz v. Bollinger*, 539 U.S. 244 (2003), was pending before the Supreme Court, much publicity was centered around the fact that the University of Michigan routinely added the equivalent of an entire letter grade to the admissions index of underrepresented minority students. An African-American student with a high school grade point average of 2.95 would thus be preferred to an Asian American student with a high school grade point average of 3.94 (just shy of straight As) all other things being equal. The *Gratz* case rejected such a formulaic approach, but it did not reject the size of the preference granted to minority students. And indeed, the evidence suggests that the size of the preference grew at the University of Michigan in the period following the *Gratz* decision. See infra at n. 39.

Michigan's affirmative action policies were not more over-the-top than those of other universities. Lawsuits filed against the University of Georgia, the University of Texas and the University of Washington prior to the Supreme Court's decision in *Gratz* brought to light similar practices. Hopwood v. Texas, 78 F.3d 932 (5th Cir. 1996), cert. denied 518 U.S. 1033 (1996)(law school); Smith v. University of Washington, 233 F. 3d. 1188 (9th Cir. 2000)(law school); Johnson v. Board of Regents, 106 F. Supp. 2d 1362 (S.D. Ga. 2000), aff'd, 263 F.3d 1234 (11th Cir. 2001)(undergraduate admissions policies). See also Robert Lerner & Althea Nagai, Racial and Ethnic Preferences in Undergraduate Admissions at Six North Carolina Public Universities, Center for Equal Opportunity (May 28, 2007)(finding similar preferences at competitive universities in North Carolina), available at http://www.ceousa.org/content/view/442/100/.

Some of the most discriminatory policies are at professional schools. At law schools, for example, the average black student has an academic index that is more than two standard deviations below that of his average white classmate. Richard Sander, A Systemic Analysis of Affirmative Action in American Law Schools, 57 Stan. L. Rev. 367, 393 (2004).

[61] In addition, many have asserted that there is a shortage of Americans trained in science and engineering and that this shortage will likely get worse. If a particular segment of the population is underrepresented in these fields, it is only prudent for colleges and universities, employers and government to look into what can be done to increase their participation. National Science Foundation, Future Scarcities of Scientists and Engineers: Problems and Solutions (1992).

[62] Richard Sander & Roger Bolus, Do Credentials Gaps in College Reduce the Number of Minority Science Graduates, Working Paper 2 (Draft July 2009)(using data from 2003)(hereinafter "Sander & Bolus").

[63] Unlike African-Americans, Hispanics in science and engineering do not appear to be underrepresented relative to Hispanics in other college disciplines, such as the humanities. Id. Relative to their initial interest, however, they are underrepresented. Ordinarily one would expect a language minority to be overrepresented in science and engineering, since those disciplines do not require the same language skills as the humanities.

[64] Elizabeth Culotta & Ann Gibbons, Minorities in Science: Two Generations of Struggle: Special Report Overview, 258 Science 1176 (November 13, 1992).

[65] Calvin Sims, What Went Wrong: Why Programs Failed, 258 Science 1185, 1185 (November 13, 1992).

[66] *Id.* at 1187.

[67] *Id.*

[68] Frederick L. Smyth & John J. McArdle, Ethnic and Gender Differences in Science Graduation at Selective Colleges with Implications for Admission Policy and College Choice, 45 Res. Higher Ed. 353, 357 (2004) (calling this finding "consistent" and citing a number of studies dating back to the late 1970s) (hereinafter "Smyth & McArdle").

[69] Alexander W. Astin & Helen S. Astin, Undergraduate Science Education: The Impact of Different College Environments on the Educational Pipeline in the Sciences 3-9, Table 3.5 (1993), available at http://www.eric.ed.gov/PDFS/ED362404.pdf (hereinafter "Astin & Astin").

[70] Rogers Elliott, A. Christopher Strenta, Russell Adair, Michael Matier & Jannah Scott, The Role of Ethnicity in Choosing and Leaving Science in Highly Selective Institutions, 37 Res. Higher Ed. 681, 692-93 (1996)(hereinafter "Elliott").

[71] Sander & Bolus, supra n. 13 at 3. Sander and Bolus also report that among the University of California students enrolling from 1992 to 2006, 52.6 percent of Asians declared an intention to major in science and engineering, as did 37.5 percent of Blacks/Hispanics and 34.7 percent of whites.

[72] Elliott, supra n. 21, at 694. See also National Science Foundation, Women, Minorities, and Persons with Disabilities in Science and Engineering (NSF Report 99-338) (1999); National Science Foundation, Future Scarcities of Scientists and Engineers: Problems and Solutions (1990)(finding persistence rates of 43 percent for majority students and 21 percent for minority students); T.L. Hilton, J. Hsia, D.G. Solorzano, & N.L. Benton, Persistence in Science of High Ability Minority Students (1989) (reporting that 54 percent of Asian, 44 percent of white, 36 percent of black and 29 percent of Latino high school seniors who had intended to attend college and major in science or engineering were doing so two years later).

[73] Smyth & McArdle, supra n. 19, at 361-63.

[74] Astin & Astin, supra n. 20, at 3-9, Table 3.5; Elliott, supra n. 21, at 694; Smyth & McArdle, supra n. 19, at 357; Sander & Bolus, supra n. 13.

[75] See id. See also William G. Bowen & Derek Bok, The Shape of the River: Long-Term Consequences of Considering Race in College and University Admissions (1998).

[76] Sadly, some of the most promising avenues for K-12 improvement are not being pursued. Consider the D.C. Opportunity Scholarship Program, which is (or more accurately was) the District of Columbia's federally-funded school voucher program, providing $7,500 in tuition per year to low-income students to attend private schools. The overwhelming majority of its beneficiaries were African-Americans or Hispanics. At the height of the program, it allowed over 1,700 students to escape the grasp of D.C.'s dysfunctional public school system and attend quality private schools.

The best hope that one day racially preferential admissions policies will be a quaint relic of the past comes from programs like it. Careful social science research, led by Dr. Patrick Wolf of the University of Arkansas College of Education and Health Professions, concluded that, after three years of study, students receiving these scholarships had improved reading skills–equivalent to 3.1 additional months of study--relative to their counterparts, who had remained in D.C. government schools. In other words, bit by bit, the achievement gap was closing.

The program enjoyed the warm support of D.C. Mayor Adrian Fenty and D.C. Public Schools Chancellor Michelle Rhee–both reform-minded Democrats. It was also supported by almost 70 percent of D.C. residents. But teachers' unions let it be known they opposed it–as they do all school choice programs. Buried in the 1,000-page spending bill for 2010, was language that closed down the program to new students. The program was effectively gutted when President Obama signed the bill into law in December.

Three months later, Senator Joseph Lieberman (I-CT), joined by co-sponsors Robert Byrd (D-WV), Susan Collins (R-ME), John Ensign (R-NV), Dianne Feinstein (D-CA), Jon Kyl (R-AZ), and George Voinovich (R-OH), led a valiant effort to revive the program. Without support from the Administration, however, their effort was defeated by a vote of 55 to 42. The Washington Scholarship Fund, which administers the D.C. Opportunity Scholarship Program, closed its doors at the end of the school year in 2010.

[77] Dr. Fortmann testified: "The root problem, as I've seen and as I say I've been working intimately with some of these people is the dearth of mathematics content knowledge among elementary teachers. It's really quite appalling and it extends to many middle school teachers as well. And until we solve that, improvement and innovations at the high school and college levels really can't have much effect. And the reason they can't have much effect is the cumulative nature of mathematics that I just mentioned." Tr. at 47-48. Dr. Fortmann also directed the Commission's attention to evidence that fifth and sixth grade teachers responsible for teaching students mathematics were themselves not competent at arithmetic. Tr. at 38-49.

[78] Elliott, supra n. 21, at 681.

[79] *Id.* Among the credentials that mattered most were number of science courses taken, average grades in high school science courses and SAT-Math score.

[80] *Id.* at 701.

[81] *Id.* at 702. This estimate, of course, was based on the assumption that the student started out with a desire to major in science or engineering. Whether a student with no particular plans to major in science or engineering

is more likely to graduate with a science or engineering degree if he attends a school at which he is properly matched is a more complex matter. As the Elliott team demonstrated, students with higher SAT Math scores are more likely to begin college with a desire to major in science. Consequently, institutions A-E likely have more students interested in pursuing science than Institutions F-K and thus would naturally be expected to award a higher proportion of science degrees, since that is what their students desire. And indeed they did. The Elliott team reported that Institution A-E were about twice as likely to award science degrees as Institutions F-K, with about 28% of the first group's bachelor's degrees being in science and about 15% of the second group's. Nevertheless, as they point out, "a 54 percent chance of getting one of the 15 percent of the degrees that are in science is nearly twice as good as a 15 percent chance of getting one of the 28 percent of degrees that are in science." Id. at 702.

[82] In theory, intensive remedial instruction is supposed to bridge the gap between the top and the bottom. But not every theory works out in reality. The educational experience at elite institutions is meant to be a full-time job and then some. With only twenty four hours in a day, something has to give. Every hour a minority student spends in a remedial classroom, sometimes struggling to stay on top of material other students are having less trouble with, is an hour other students can spend getting a deeper understanding of that material. The game of catch-up is thus never-ending.

[83] Elizabeth Culotta, Black Colleges Cultivate Scientists, 258 Science 1216 (November 13, 1992)(hereinafter "Culotta").

[84] The Elliott team members were particularly impressed that HBCUs are able to graduate large numbers of students in science and engineering despite entering credentials that were significantly lower than those ordinarily found at elite institutions. Students at Xavier University, for example, were reported to have SAT Math scores averaging around 400, yet half of the class was majoring in science. If elite schools could do the same with minority students (or with students in the bottom third of the class generally), it would be astonishing. In fact they do the opposite. They are able to award far fewer science or engineering degrees to African-Americans than one would expect given the number of African-American students in their classes. Elliott, supra n. 21, at 700.

[85] Cullotta, supra n. 34, at 1218.

[86] Indeed accreditation authorities have often demanded it. See, e.g., Gail Heriot, Affirmative Action in Law Schools, 17 J. Contemp. Legal Issues 237 (2008)(symposium issue)(discussing the American Bar Association diversity requirements for law schools and its efforts to pressure law schools into increasing minority representation). See also Susan Welch & John Gruhl, Affirmative Action and Minority Enrollments in Medical Schools and Law Schools 80 (1998).

[87] 539 U.S. 306 (2003).

[88] Althea K. Nagai, Racial and Ethnic Preferences in Undergraduate Admission at the University of Michigan, Center for Equal Opportunity (October 17, 2006), available at http://www.ceousa.org/content/blogcategory/78/100/. See also Fisher v. University of Texas, 645 F. Supp. 2d 587 (W.D. Tex. 2009).

[89] Russell K. Nieli, The Changing Shape of the River: Affirmative Action and Recent Social Science Research, 17 Academic Questions 7 (2004).

[90] Stephen Cole & Elinor Barber, Increasing Faculty Diversity: The Occupational Choices of High Achieving Minority Students 124, 212 (2003)(citations omitted)(emphasis supplied).

[91] Robin Wilson, The Unintended Consequences of Affirmative Action, The Chronicle of Higher Education 10 (January 31, 2003).

[92] *Id.*

[93] Smyth & McArdle, supra n. 19, at 373.

[94] *Id.* at 373.

[95] Sander & Bolus, supra n. 13.

[96] Sander & Bolus, supra n. 13, at 14-20.

[97] Sander & Bolus, supra n. 13, at 20-23.

[98] Sander & Bolus, supra n. 13, at 23-24.

[99] The figures for law schools grades are available and particularly instructive: In elite law schools, 51.6 percent of African-American law students have first-year GPAs in the bottom 10 percent of their class as opposed to only 5.6 percent of white students. Nearly identical gaps exist at law schools at all levels (with the exception of historically minority schools). At mid-range public schools, the median African-American student's first-year grades corresponded to the 5$^{th}$ percentile among white students. For mid-range private schools it was 7$^{th}$. With disappointingly few exceptions, African-American students were grouped towards the bottom of their class. Moreover, contrary to popular belief, the gap in grades did not close as students continued through law school.

Instead, by graduation, it had gotten wider. Richard Sander, A Systemic Analysis of Affirmative Action in American Law Schools, 57 Stan. L. Rev. 367, 427-36, Tables 5.1, 5.3 & 5.4 (2004). I am not aware of anyone who disputes these figures, and indeed some critics of Sander's work appear to have conceded their accuracy. See Ian Ayres & Richard Brooks, Does Affirmative Action Reduce the Number of Black Lawyers?, 57 Stan. L. Rev. 1807, 1807 (2005) ("Richard Sander's study of affirmative action at U.S. law schools highlights a real and serious problem: the average black law student's grades are startlingly low").

[100] William G. Bowen & Derek Bok, The Shape of the River: Long-Term Consequences of Considering Race in College and University Admissions 72 (1998).

[101] United States Commission on Civil Rights, STEM Briefing Transcript, p. 146 (Commissioner Yaki speaking).

[102] *Id.* p. 34.

[103] Richard H. Sander, *A Systemic Analysis ofAffirmative Action in American Law Schools*, 57 STAN. L. REV. 367 (2004).

[104] *See, e.g.*, Ian Ayers & Richard Brooks, *Does Affirmative Action Reduce the Number ofBlack Lawyers?*, 57 STAN. L. REV. 1807 (2005); David B. Wilkins, *A Systemic Response to Systemic Disadvantage: A Response to Sander*, 57 Stan. L. Rev. 1915 (2005); Daniel E. Ho, Scholarship Comment, *Why Affirmative Action Does Not Cause Black Students To Fail the Bar*, 114 YALE L.J. 1997 (2005).

[105] Richard H. Sander, *A Reply to Critics*, 57 STAN. L. REV. 1964 (2005); Richard H. Sander, *Mismeasuring the Mismatch: A Response to Ho*, 114 YALE L. REV. 2005 (2005).

[106] *See, e.g.*, David L. Chambers, Timothy T. Clydesdale, William Kidder & Richard Lempert, *The Real Impact ofEliminating Affirmative Action in American Law Schools: An Empirical Critique ofRichard Sander's Study*, MICH. L & ECON WORKING PAPER NO. 05-007 (2005), *available at* http://papers.ssrn.com/sol3/papers.cfm?abstract_id=730506##; David L. Chambers, Timothy T. Clydesdale, William Kidder & Richard Lempert, *Affirmative Action in American Law Schools: A Critical Response to Richard Sander's "A Reply to Critics"*, MICH. L & ECON WORKING PAPER NO. 06-001, *available at* http://papers.ssrn.com/sol3/papers.cfm?abstract_id=886382; Katherine Y. Barnes, *Is Affirmative Action Responsible for the Achievement Gap Between Black and White Law Students?*, 101 NW U. L.REV. 1759 (2007).

[107] Richard H. Sander & Jane Yakowitz, *The Secret ofMy Success: How Status, Prestige and School Performance Shape Legal Careers*, (2010), *available at* http://papers.ssrn.com/ sol3/papers.cfm?abstract_id=1640058.

[108] Admittedly, the exact nature and extent of this "mismatch" effect has been obscured by resort to vague language in the recommendations, such as "large deficit." It should be noted that initial approval of this report's recommendations was scuttled when a majority of Commissioners were unwilling to endorse this imprecise language. *See* USCCR Business Meeting Transcript June 11, 2010, pp. 27-34.

[109] Rogers Elliott, A. Christopher Strenta, Russell Adair, Michael Matier & Jannah Scott, The Role of Ethnicity in Choosing and Leaving Science in Highly Selective Institutions, 37 Res. Higher Ed. 681 (1996).

[110] Richard Sander & Roger Bolus, Do Credentials Gaps in College Reduce the Number of Minority Science Graduates?, Working Paper (Draft July 2009).

[111] Frederick L. Smyth & John McArdle, Ethnic and Gender Differences in Science Graduation at Selective Colleges with Implications for Admission Policy and College Choice, 45 Res. Higher Ed. 353 (2004).

[112] It should be noted that the work of these three investigative teams does not come out of nowhere. It is consistent with a long history of research that dates back to the 1960s. That literature includes the work of Elinor Barber and Stephen Cole, which found that affirmative action beneficiaries at elite schools are less likely than similarly-credentialed minority students attending less elite schools to desire to become college professors. See Elinor Barber & Stephen Cole, Increasing Faculty Diversity: The Occupational Choices of High Achieving Minority Students (2003). It also includes a wealth of literature like James A. Davis's The Campus as A Frog Pond: An Application of the Theory of Relative Deprivation to Career Decisions of College Men, which does not deal with race-conscious admissions practices directly, but which nevertheless sheds light on the processes at work at schools where those practices are employed. Davis found that, for better or worse, among the college students he studied, students who get good grades at less competitive schools will think more highly of their own academic skills than otherwise identical students who attend more competitive schools and thus earn lower grades. He further found that good self-opinion influences career choices. James A. Davis, The Campus as a Frog Pond: An Application of the Theory of Relative Deprivation to Career Decisions of College Men, 72 Am. J. Socio. 17, 30-31 (1966)("[T]hese ideas ... challenge the notion that getting into the 'best possible' school is the most efficient route to occupational mobility. Counselors and parents might well consider the drawbacks as well as the advantages of sending a boy to a 'fine' college, if, when doing so, it is fairly certain he will end up in the bottom ranks of his graduating class."). See also Marjorie Seaton, Herbert W. Marsh & Rhona G. Craven, Big-Fish-Little-Pond Effect: Generalizability and Moderation: Two Sides of the Same Coin,

47 Am. Educ. Res. J. 390 (2010); Herbert W. Marsh, Ulrich Trautwein, Oliver Ludke, Jurgen Baumert & Olaf Koller, The BigFish-Little-Pond Effect: Persistent Negative Effects of Selective High Schools on Self-Concept After Graduation, 44 Am. Educ. Res. J. 631 (2007); Herbert W. Marsh, Chit-Kwong Kong & Kit-Tai Hau, Longitudinal Multilevel Models of the Big Fish Little Pond Effect on Academic Self-Concept: Counterbalancing Contrast and Reflected Glory Effects in Hong Kong Schools, 78 J. Personality & Soc. Psych. 337 (2000).

[113] Commissioners Yaki and Melendez are less cautious. Three years ago, in connection with the Commission's Report on Affirmative Action in Law Schools, they were utterly disdainful of the mismatch theory and quite certain that the issue was the invention of sinister forces. They asserted "[t]he proof that affirmative action works in law schools is overwhelming" and that "the success of the policy" is "unmistakable." See U.S. Commission on Civil Rights, Briefing Report: Affirmative Action in Law Schools, Joint Dissent of Commissioner Arlan D. Melendez and Commissioner Michael Yaki at 188 (2007). So certain were they that mismatch was not an issue in law school or anywhere else that they dissented even from the Commission's recommendation that more research be conducted—hardly the cautious course of action. Now they argue with equal certainty that "[t]his effect has been exaggerated" in the science and engineering context. This statement is curious given that the current evidence in the science and engineering context points in only one direction. In the future, it could always turn out that researchers were mistaken about the mismatch effect. Or it could turn out that the effect was larger or smaller than they now believe it to be. But at present there is no basis upon which to conclude that it "has been exaggerated." To baldly assert otherwise is to substitute swaggering certitude for rational analysis.

[114] As far as I know, no one has ever made such an argument in the context of products liability. In that area, for example, a manufacturer can be held liable if it fails to warn of a "foreseeable" risk. See Restatement (Third) of Torts: Products Liability section 2. To exempt manufacturers from liability in cases in which the evidence of risk was not yet utterly conclusive would be an extraordinary (and in my view inappropriate) change of policy. A large proportion of the warnings on pharmaceutical products are for risks that have not yet been conclusively proven to exist. Science, like life, is full of uncertainties. In this case, the risk is not only foreseeable, it has actually been foreseen by a number of commentators of which Drs. Elliott, Strenta, Adair, Matier, Scott, Smyth, McArdle, Sander, and Bolus are just a few. Moreover, it is not simply foreseen, it has been tested empirically and strong evidence of its existence has been found in three independent studies.

[115] Interestingly, rather than discuss that evidence or indeed anything directly related to the current report, the dissenting commissioners prefer to discuss an earlier Commission report—one in which the issue was the effect of affirmative action in law schools rather than its effect on science and engineering students at the undergraduate level. See United States Commission on Civil Rights, Report on Affirmative Action in Law Schools (2007). In that report, the Commission examined research by Dr. Richard Sander that appears to show that mismatched law students are less likely to complete law school and pass the bar than are identically-credentialed students who attend law schools at which they are not mismatched. See Richard Sander, A Systemic Analysis of Affirmative Action in American Law Schools, 57 Stan. L. Rev. 367 (2004). Commissioners Yaki and Melendez take the position that this research has been "soundly debunked in a sustained manner." I can only imagine that they have not read the articles they cite. For the most part, I will rely on my analysis in my Commissioner's Statement to the Report on Affirmative Action in Law Schools Report, since I have already dealt with most of the criticisms they cite. See also Gail Heriot, Affirmative Action in Law Schools, 17 J. Contemp. Legal Issues 237 (2008)(another version of same essay).

A few points should nevertheless be made here either because they are new or in need of repetition to give the reader of this report a taste of the evidence examined in the earlier report:

1) The evidence behind Sander's conclusions in his law school study is not yet conclusive. Nor does the Commission report claim that it is. More research needs to be done and that is a significant part of what the Commission recommended. Some of the dissenting commissioners' citations, however, are quite misplaced. For example, the article by David Wilkins does not dispute Sander's conclusion that there are today fewer, rather than more, African-American attorneys as a result of racially preferential admissions policies. Instead, Wilkins argues that even assuming mismatched African-Americans leave law school or never pass the bar at higher rates than they would under race-neutral admissions policies, those who do make it get the benefit of having the opportunity to network with more elite law students. David B. Wilkins, A Systemic Response to Systemic Disadvantage: A Response to Sander, 57 Stan. L. Rev. 1915 (2005).
2) The dissenting commissioners cite Katherine Y. Barnes, Is Affirmative Action Responsible for the Achievement Gap Between Black and White Students?, 101 Nw. U. L. Rev. 1759 (2007), as part of a

second wave of criticism of the Sander study. Scholars, however, have been unable to replicate Dr. Barnes' results. See Doug Williams, Does Affirmative Action Create Educational Mismatches in Law School, Draft at 9 (January, 2010)(presented at the 20th Annual Conference of the American Law and Economics Association). Efforts to do so by Williams "produce[d] results generally consistent with the mismatch hypothesis." Id. Indeed, Dr. Barnes' own efforts to replicate her original findings have produced very different results from those in the original article—results that provide some support for Sander. For example, in her original article she reported that students in the 5th percentile nationally for entering academic credentials passed the bar after attending a historically black law school at a rate of only 12.9%. Her revised is 77.9%. In a correction and update that has been submitted to the Northwestern University Law Review, she reports that her new results do indeed indicate that HBCUs "boost graduation and bar passage rates for students with low credentials"—although she is not yet willing to concede that the total pattern of results supports mismatch. Katherine Y. Barnes, Correction and Update to Is Affirmative Action Responsible For The Achievement Gap Between Black And White Law Students?, forthcoming in the Northwestern University Law Review. Dr. Barnes agrees with the Commission that more research is necessary and that the Bar Passage Study database is not adequate for that research.

3) The Williams article provides general support for Sander. "Although the results here are not conclusive," he writes, "I find much more evidence for mismatch effects than previous research, which has been dismissive of the mismatch hypothesis." Significantly, Williams goes on to explain that the Bar Passage Study, from which the data for Sander's and his studies were obtained, does not categorize law schools by academic tier. Consequently, Williams points out, it is difficult to find evidence of mismatch even if a serious mismatch problem really is there. Williams urges that further studies be conducted on bar passage data from large states like California, Florida or Texas so that the mismatch theory in law schools can be confirmed or refuted Williams at 34. Williams' recommendation is thus in accord with the Commission's.

4) Surely all fair-minded scholars would agree that obtaining data with which to confirm or refute Sander's conclusions concerning law schools is worthwhile and important. But if one were to expect those who critiqued the original study uniformly to be fair-minded scholars, one would be disappointed. William Kidder, one of the team of David Chambers, Timothy Clydesdale, William Kidder & Richard Lempert, whose critique of the Sander study is cited by Commissioners Yaki and Melendez in their Footnote 6, wrote to the State Bar of California's Committee of Bar Examiners urging that the committee deny Sander and his co-authors access to the California Bar Examination data—probably the richest and best source available upon which to test the hypothesis further. Among other things, Kidder argue that disclosure of the data "risks stigmatizing African-American attorneys regardless of how successful they may be in legal practice." Sander, of course, had not requested names or any information that would allow him to identify particular persons. At the time he coauthored the article cited by the dissenters, Kidder was employed as a researcher with the Equal Justice Society, an organization that describes its mission as to "marshal our forces to defeat the right wing assault on social and racial justice." More recently, he has been employed as an administrative staff member at the University of California. See United States Commission on Civil Rights, Report on Affirmative Action in Law Schools, Statement of Commissioner Gail Heriot at 148 (2007). Similarly, David Chambers was the former president of the Society of American Law Teachers, which also wrote a letter to the State Bar of California opposing scholarly access to the data. That letter raised the specter of litigation if the data is disclosed. Id.

5) Commissioners Yaki and Melendez seem to suggest that there is something strange about the fact that the word "mismatch" does not appear in Dr. Sander's most recent article. Richard H. Sander & Jane Yakowitz, The Secret of My Success: How Status, Prestige and School Performance Shape Legal Careers, Draft (August 9, 2010), available athttp://papers.ssrn.com/sol3/papers.cfm? abstract_id= 1640058. The reason for this choice of words is easy to explain: The paper isn't about mismatch, but rather about how law school grades have more impact on future success in the legal profession than the prestige of the law school a lawyer attended (controlling for other factors like LSAT). The study did not make a distinction between students who were mismatched and those who were not. The fact that the word "mismatch" was not used is thus no more significant than the absence of the word "elephant" from the article.

116 Dissent Statement of Yaki and Melendez at 1; Transcript at 146.

117 My own view is that it is naïve to expect the Commission staff to please all Commissioners as to "balance" when they invite experts to testify at a briefing. No matter how well-balanced the panel is or is not, someone will

complain. Any procedure that places the responsibility for balance on the Commission staff's shoulders is thus begging for controversy. Instead, that responsibility should be on the Commissioners, each one having the right to call an expert witness to a briefing (if he or she feels the panel is unbalanced). That way if a panel does indeed fail to reflect the spectrum of responsible opinion on an issue, the Commissioners have no one to blame but themselves.

Sadly, it has been observed by the Commission staff that when witnesses generally congenial to Commission Yaki's point of view agree to testify, they sometimes mysteriously withdraw shortly after their identities are made available to the members of the Commission. Commissioner Yaki can then berate the staff for failing to properly balance the panel and complain that the Commission's report is somehow illegitimate. Again, the solution is to alter the Commission's procedures. If the Commissioners were responsible for identifying and securing witnesses, there would be no occasion for concerns of this type.

It is well worth noting that Congress does not charge some "neutral" staff with the responsibility for balancing opinion at Congressional hearings. Instead, the majority chooses a number of witnesses and the minority is permitted to choose a somewhat smaller number. Under the circumstances, neither party can complain about whether the full spectrum of opinion was reflected in the panel. Instead, disagreements are limited to the more mundane question of whether the minority has been accorded the right to call an adequate number of witnesses. I would amend our procedures to conform to Congressional practice. I believe this would eliminate some of the controversy over panel balance that periodically rears its head at our briefings.

[118] Also contacted for advice on speakers was Peter Henderson, Director on the Board of Higher Education and Workforce of the National Academy of Sciences and Sigal Alon, Senior Lecturer in the Department of Sociology at Tel Aviv University.

[119] According to the magazine Science, Dr. Tapia's graduate program in mathematics at Rice University has graduated far greater than the average number of underrepresented minorities than is the national average, causing the National Research Council to cite his program's success in one of its reports. Paul Selvin, Math Education: Multiplying the Meager Numbers, 258 Science 1200, 1201 (November 13, 1992).

[120] STEM Testimony at 146.

[121] Meeting of November 20, 2009, Transcript at 8-9. See Rebuttal Statement of Commissioner Gail Heriot to Report on Historically Black Colleges and Universities (issued in conjunction with this report) at n.1 (cataloguing occasions upon which Commissioner Yaki has walked out of ordinary business meetings to defeat a quorum or joined a meeting only when a quorum is already established.)

[122] *See* 42 U.S.C. 1975(b): "The Commission shall be composed of eight members." The subsection then goes on to describe how members of the Commission will be appointed by the President and Congress. But it is silent on the subject of staff.

# INDEX

## #



## A



## B



## C

## D

## E



## F



## G

## H



## I



## J



## K



## L

## M



## N



## O

## P



## Q



## R

## S



## T

## U



## V



## W



## Y